DISASTROUS FLOODS

AND THE

DEMISE OF STEEL IN

JOHNSTOWN

DISASTROUS FLOODS
AND THE
DEMISE OF STEEL IN
JOHNSTOWN

PAT FARABAUGH

FOREWORD BY RICHARD BURKERT
PRESIDENT AND CEO OF JOHNSTOWN AREA HERITAGE ASSOCIATION

THE
History
PRESS

Published by The History Press
Charleston, SC
www.historypress.com

Front and back cover images courtesy of the Johnstown Area Heritage Association.

First published 2021

Manufactured in the United States

ISBN 9781467150019

Library of Congress Control Number: 2021943564

This book is dedicated to the victims of the 1889, 1936 and 1977 floods in Johnstown, as well as the men and women who worked for the Cambria Iron Works, Cambria Steel, Midvale Steel and Bethlehem Steel in the city and surrounding area.

CONTENTS

FOREWORD

I became aware of Pat Farabaugh's work in the history of Johnstown, Pennsylvania, when he presented an early draft of *Disastrous Floods and the Demise of Steel in Johnstown* at a Pennsylvania Historical Association conference. I am delighted that the author has seen the manuscript through to publication. The book will instantly become the only history of Johnstown's steel industry for a general audience. Farabaugh makes this complex story available in a readable, coherent narrative—one that assigns responsibility for the decline and disappearance of this colossal enterprise. Farabaugh is an excellent writer, and his use of first-person and newspaper accounts brings an immediacy to his narrative.

Two themes run in a parallel course through *Disastrous Floods and the Demise of Steel in Johnstown*—Johnstown's major floods and the rise and decline of Johnstown's iron and steel industry.

Johnstown's identity as a steel center began with the founding of the Cambria Iron Company in 1854. A technological leader blessed with abundant natural resources nearby, Johnstown was, for a time, the dominant steel producer in the early era of the industry. While Pittsburgh and other communities surpassed Johnstown's steel output, Johnstown remained a major steel center well into the twentieth century. In 1923, the Johnstown plant became a part of the Bethlehem Steel Corporation—the country's second largest steel producer.

Responses varied after each of Johnstown's three major floods. The May 1889 flood was caused by the breaking of a dam that held back a lake owned by a group of wealthy industrial financiers from Pittsburgh. Despite immense destruction and the loss of 2,200 lives, the community began recovery immediately after the disaster. In 1890, Johnstown had a population of 21,805; thirty years later, it stood at a peak of 67,327. The output of the steel mills quadrupled in the two decades after the Great Flood.

The St. Patrick's Day flood of 1936 delivered a devastating blow to the local economy during the Great Depression, and residents began to demand that something be done to end the flooding menace. On August 13, 1936, President Franklin Roosevelt promised flood control for Johnstown and nine miles of Johnstown's rivers were channelized. Johnstown declared itself "flood-free." The onset of World War II pushed steel production to new heights, and in the post–World War II period, newly organized steelworkers entered the middle class. A steel strike in 1959 resulted in changes to work rules in the mills. Farabaugh argues that these changes played an important role in the plant's later decline.

The July 1977 flood was caused by a record rainfall in the Johnstown area—up to twelve inches in six hours. Quickly, 7,300 steelworkers lost their jobs. Soon after the 1977 flood, Bethlehem Steel canceled the installation of a partially completed basic oxygen furnace, but it installed two massive electric arc steel furnaces—the most powerful in the country. But Bethlehem's losses did not allow the firm to also install a continuous caster, the other piece of modern technology needed to make the plant competitive. While the demise of Johnstown's steel industry became apparent after the 1977 flood, Farabaugh argues that the unanticipated disaster was not the cause of the rapid downward spiral. In 1982, unemployment in Johnstown reached 17.3 percent—the highest rate in Pennsylvania. In July 1992, Bethlehem Steel's Johnstown plant ceased operations, except for the Gautier Plant and the Wire Mill, which were acquired by other operators. In 2001, the corporation filed for bankruptcy.

Farabaugh cites mismanagement, international competition, reluctance to invest in capital improvements and unsustainable wages as the causes of Bethlehem's decline, rather than the 1977 flood.

Farabaugh is himself a descendent of industrial workers and writes with a familiarity and pride about this heritage. In a note, he credits a source as being his grandfather John Earl Boland, who worked as a laborer in Cambria County coal mines from 1932 until his retirement in 1970. Farabaugh's use of interviews as a source of information, along with newspaper accounts, provides the work with a vitality that enhances the story's authenticity.

It is amazing to me how the heavy imprint of industry can vanish in a matter of decades. The local coal industry began its final decline in the 1960s, and now, even the colossal Franklin Works—the massive plant that was the center of steelmaking in Johnstown since 1901—has been turned to scrap. Readers of *Disastrous Floods and the Demise of Steel in Johnstown* will come away with a better understanding of how something that appeared so permanent—Bethlehem's Johnstown plant—could vanish after 140 years of successful operation.

Richard Burkert
President and CEO
Johnstown Area Heritage Association

ACKNOWLEDGEMENTS

First and foremost, I praise and thank God, who has granted me countless blessings and opportunities. I thank my wonderful wife, Jenna, for her support throughout this project, and I am appreciative of her proofreading expertise. I owe a debt of gratitude to my mother, Betty Ann, for taking me to the library as a kid and instilling in me a passion for history and the written word.

I am grateful to Richard Burkert, president and CEO of the Johnstown Area Heritage Association, for sharing his vast knowledge of the Johnstown region and its history, as well as his willingness to proofread the book and write a foreword to it. My gratitude is also extended to JAHA historian Amy Regan and former JAHA archives and collections specialist Marica Kelly, who helped collect many of the photographs that are included in the book. Special thanks also go to Conrad Suppes, a former member of Boy Scout Troop 2183, who completed an oral history of the 1977 flood for his Eagle Scout project. His project is titled "Documenting the Stories of the July 20, 1977 Flood," and his interviews proved invaluable.

Thanks also go to all those involved with another oral history project, one titled "In the Age of Steel: Oral Histories from Bethlehem, Pennsylvania." Interviews for this project, overseen by Julia Maserjian, were conducted by students and faculty at Lehigh University from 1974 to 1977.

I want to thank Noah Brunhuber and Amber Elliott, two of my students at Saint Francis University, for their assistance in transcribing interviews. Thanks also to many of my colleagues at Saint Francis for their support throughout the research and writing of the book. I work at a special place with special people.

Among these special people is Saint Francis library assistant Leslie Conrad. Leslie tracked down and scanned some of the photographs in the book. This is the second book project that she has helped me with, and I am, once again, grateful for her assistance. There are also several photographs in the book that were provided by the National Museum of Industrial History in Bethlehem, Pennsylvania. Andria Zaia, curator of collections at this museum, helped find these photographs, and my appreciation is extended to her as well.

I owe a debt of gratitude to the staff members at the Cambria County Library in downtown Johnstown for their help as I combed through newspaper archives. Thanks also to Linda Ries, editor of the *Pennsylvania History Journal*, for providing valuable input and clarification on information related to the 1977 flood. I also appreciate the feedback I received from different members of the Pennsylvania Historical Association. These folks have a passion for the history of our commonwealth, and I am grateful for their insight.

I am also grateful for all of the help provided by the talented team at The History Press. Writing a book is a marathon, and it is easy to run out of gas before you get to the end of the race. The group at this publishing house helped me get to the finish line. Thank you to all involved, with a special thanks to Ashley Hill, copy editor; J. Banks Smither, acquisitions editor; Jenni Tyler, publicist; and Crystal Murray, senior sales specialist.

INTRODUCTION

In 1889, Johnstown, Pennsylvania, suffered a catastrophic flood of biblical proportions, gaining the attention of the world. Volumes of works have been written about the 1889 Johnstown flood, including renowned historian David McCullough's seminal record of the event, *The Johnstown Flood: The Incredible Story Behind One of the Most Devastating Disasters America Has Ever Known*. For many years, the "Great Flood" was considered one of the three worst natural disasters in American history, along with the 1900 Galveston hurricane and the 1906 San Francisco earthquake and fire. The 1889 deluge killed 2,209 people, including 99 entire families and 396 children; 1 out of every 9 people in Johnstown was killed. Roughly one-third of the dead were never identified by authorities, and four square miles of the city's downtown and more than 1,600 homes were destroyed.[1]

Another flood, albeit one far less devastating, struck Johnstown in March 1936, killing twenty-four people in the city and surrounding region and destroying seventy-seven buildings. Five months after this flood, on August 13, 1936, President Franklin D. Roosevelt visited the city. He authorized the U.S. Army Corps of Engineers to channelize the rivers where they flowed through Johnstown at a cost of $8.7 million. The goal of this "Local Flood Protection Project," which included the construction of "flood protection walls," was to increase the rivers' capacity to prevent future flooding. On August 28, 1937, Roosevelt signed the Omnibus Flood Control Act. The bill stipulated that "flood protection shall be provided for said city [Johnstown] by channel enlargement or other works." Roosevelt told Johnstowners: "We want to keep you from having these floods again.

Downtown Johnstown following the Great Flood of 1889. *Johnstown Area Heritage Association.*

The federal government, if I have anything to do with it, will cooperate with your state and community to prevent further flooding."[2]

Shortly after Roosevelt signed this bill, James Bogardus, Pennsylvania's secretary of forest and waters, recommended that three reservoirs be built within the tributary system of Johnstown's Stonycreek River in addition to the proposed channel system, but this initiative was not pursued. The Pittsburgh Army Corps of Engineers announced plans for a reservoir along the Stonycreek, but this plan was also scrapped due to cost, as well as the risk that it could pose to the many active underground coal mines near the river. Work on the channel system began in August 1938 and was completed five years later. The channelization project was designed to prevent a flood equal to the 1936 deluge.[3]

Following completion of the channel system, Roosevelt sent a letter to Walter Krebs, chairman of a group called the Flood-Free Johnstown Committee. The president told Krebs, "Johnstown, from now onwards, will be free from the menace of floods. Happily, for the future of Johnstown, its citizens can now devote all their energies to their ordinary pursuits without worry over the impending hazard of uncontrolled waters." Colonel Gilbert Van Wilkes, chief of the Pittsburgh Army Corps of Engineers, and Johnstown mayor Daniel Shields echoed Roosevelt's sentiment. Van Wilkes, who helped oversee the channel project, said, "We believe that the flood troubles of Johnstown are at an end. We thank the people of the city for their cooperation and salute the flood-free city of Johnstown." Shields also believed that deficiencies had been sufficiently addressed: "Johnstown looks forward with confidence to a future in which the tragedies of the river will be only a memory. Real, effective flood control will be an accomplished fact."[4]

It wasn't. On July 19, 1977, it began raining very heavily in Johnstown around 6:30 p.m., and the storm didn't relent until around 4:00 a.m. the following morning. The causal factors of this flood were different from those of the floods in 1889 and 1936. The primary cause of the 1889 flood was the failure of the South Fork Dam. The 1936 flood was the result of a large weather system of heavy rains that extended across much of the northeastern United States. The 1977 flood was the result of extremely powerful and localized thunderstorms, coupled with the collapse of the Laurel Run Dam.

The 1977 flood claimed the lives of eighty-six people. Following this third deluge, the federal government again sent money to the region: $75 million was provided for rebuilding projects, $25 million for flood-control infrastructure and $90 million for low-interest loans. As of the publication

President Franklin D. Roosevelt visited Johnstown five months after the 1936 St. Patrick's Day flood. *Johnstown Area Heritage Association.*

The 1977 flood left many streets, including Fairfield Avenue, looking more like river channels than roadways. *Johnstown Area Heritage Association.*

of this book, the city has been spared a fourth flooding disaster. Whether Johnstown is permanently "flood-free," however, is anyone's guess.

This book shares the story of these three floods while also chronicling the ebbs and flows of the steel industry in the city in the valley. The two are uniquely intertwined, and any history of one would not be complete without a close examination of the other.

Throughout most of the twentieth century, Bethlehem Steel was the main player in the steel industry in Johnstown. The history of steel production in the city dates back almost to its founding in 1800. During Johnstown's infancy and through its adolescent years, the Cambria Iron Works dominated the industry in the region as well as across the nation. "There is no question about the importance of the old Cambria Iron Works," wrote McCullough. "The age of steel in America can fairly be said to have begun there."

During a long run of success through the nineteenth century and much of the twentieth century, Johnstown's steel industry became synonymous with the city. Business boomed in the valley, and money flowed like water into the pockets of, first, Cambria Iron Works's executives and, later, Bethlehem Steel officials.

Eventually, however, the spigot dried up. By the second half of the 1970s, steel's foothold in Johnstown and across the nation had given way. In 1982, Bethlehem Steel piled up a staggering $1.47 billion in losses. Between 1981 and 1985, Bethlehem cut its workforce by more than half. By 1990, the steel giant that had once employed 164,000 employees and ranked as the ninth-largest company on the Fortune 500 list had slashed its total workforce to less than 30,000.[5]

Given that the death knell for Bethlehem Steel's operations in Johnstown sounded around the same time as the 1977 flood, it is tempting to suggest that this third major deluge to hit the city precipitated the company's demise in the region. This, however, is not supported by the evidence. The 1977 Johnstown flood was not a catalyst that triggered Bethlehem's downward spiral. While it accelerated the end of the company's operations in the city, Bethlehem had been declining well before the floodwaters rose for a third time in July 1977. The reasons for this decline are contained in the pages of this book.

A memorial stone in Johnstown lists the names of the eighty-six victims of the 1977 flood. It also includes a Bible passage: "A people who have walked

The steel industry thrived in Johnstown for decades before it declined quickly in the years following the city's third major flood. *Johnstown Area Heritage Association.*

in darkness have seen a great light: to them who live in the region of the shadow of death, a light has risen."[6]

In the roughly half century since a third major flood wreaked havoc on the city, Johnstown business and political leaders have repeatedly tried to reimagine the region's economy. And while these individuals, as well as the residents of the city, have demonstrated remarkable resiliency, the specter of the floods and the city's abandoned mills continue to hover over Johnstown ominously. They are harbingers from the past that the city seems unable to erase from its consciousness.

1
THE GREAT FLOOD

The story is as simple as it is sorrowful. It has been told in every awful and heroic detail and is now familiar in every household. If experience did not prove the probability of the situation, it would be incredible that great communities could live quietly in the immediate presence of an inconceivable possible disaster, which yet could be readily averted, and take no steps to secure the common safety. But familiarity with such possibility seems often to paralyze apprehension.…It is impossible that this event should not produce an effective determination that such disasters shall be rendered largely impossible hereafter.…Its causes are perfectly comprehended; they are entirely avoidable; and a disaster of the same kind anywhere in any degree, after this appalling warning, would be not only a calamity but a disgrace.[1]
—Quote from Harper's Weekly, *June 15, 1889*

On February 22, 1889, President Grover Cleveland signed a bill that admitted four new states—North and South Dakota, Montana and Washington—into the union two weeks before he turned the office over to his successor, James Harrison. Three months later, America's second state would suffer a "natural" disaster the likes of which the nation had never seen.

Pennsylvania was admitted into the union on December 12, 1787—five days after Delaware. Seventeen years earlier, the community of Johnstown had been established in the southwestern section of what would become known as the Keystone State. The town's first White settlers—siblings Samuel, Solomon and Rachael Adams—had relocated to the valley from the nearby community of Bedford. Johnstown was organized at the turn of the century by Josef Schantz, and the town was initially known as "Schantzstadt." An Amish farmer, Schantz arrived in Philadelphia from Switzerland in 1769 before venturing westward. With his wife and four children, he staked claim to thirty acres near the Stonycreek River. He called the community "Conemaugh," naming it after the Natives who had occupied the region. By 1820, two hundred residents lived in the village, and in 1834, the city council renamed the community "Johnstown."[8]

Sixty miles east of Pittsburgh, Johnstown sits at an elevation roughly 1,200 feet above sea level. The city rests at the bottom of a narrow valley, with steep mountains on all sides. It is enclosed by the Stonycreek and Little Conemaugh Rivers, which meet to form the Conemaugh River, one of the primary tributaries of the Allegheny River. Water from the Conemaugh River eventually finds its way to the Gulf of Mexico. The Stonycreek River originates near the community of Berlin and drains the area south of Johnstown known as the Stonycreek Valley. The Little Conemaugh River begins north of Johnstown, near the communities of Cresson and Ebensburg, and drains the Conemaugh Valley west of the city.

Johnstown's flood history dates back to at least the early nineteenth century, with twenty-three flooding events recorded between 1808 and 1937. Those who settled in the region recalled births, deaths, marriages and other significant life events based on their proximity to when the valley flooded. The area around Johnstown was very inviting to farmers—streams flowed abundantly near the confluence of the rivers, making the surrounding soil fertile. The region's water supply powered sawmills that began springing up around the valley not long after the Schantz family's arrival.[9]

During the first half of the nineteenth century, the city served as an important cargo transfer point in the Pennsylvania Main Line Canal System. The Pennsylvania legislature approved $300,000 in funding for the project, and a canal bed in Johnstown was completed. This canal route—which was built in hopes of luring business away from New York's Erie Canal System—stretched from Philadelphia to Pittsburgh. In the mountainous region of the Alleghenies, between Hollidaysburg and Johnstown in the west-central part of the state, however, canal transport was impossible due to the rugged terrain. The two sections of the canal system were connected between Hollidaysburg and Johnstown by the Allegheny-Portage Railroad. Many initially doubted the feasibility of this interchange idea, but a team of engineers led by Sylvester Welch constructed a series of twenty alternating levels and planes to regularize ascent and descent over

German immigrant farmer Josef Schantz, the "founder of Johnstown," laid out plans for the streets throughout the town. The original property he purchased is located in the community of Berlin in Somerset County. *Johnstown Area Heritage Association.*

thirty-seven miles of mountains. A double-track railroad was built across the peaks of the Alleghenies, one that is still in operation today.[10]

Included in this "canal connector" was the first railroad tunnel in the United States. Twenty feet high and twenty feet wide, with two sets of tracks running through it, the Staple Bend Tunnel above the Little Conemaugh River cut through nine hundred feet of coal, sandstone and shale. Among those who traveled through this tunnel were President William Henry Harrison, showman P.T. Barnum and author Charles Dickens. "It was very pretty, traveling thus, at a rapid pace along the heights of the mountain in a keen wind, to look down into a valley full of light and softness," noted Dickens.[11]

Work on the Staple Bend Tunnel began in November 1831, and excavation was completed in April 1833. *Photograph by author.*

In Johnstown, the canal system met the rails. Canal transportation between Pittsburgh and Johnstown had begun in 1831, and the Allegheny-Portage Railroad was completed three years later. In 1833, sixty canal boats carried 1,138 tons of freight through Johnstown; once the rail system was completed in 1834, annual tonnage through the city jumped to 5,600. For the next two decades, these rails that connected Johnstown and Hollidaysburg served as a vital link in cross-state transportation. The coal, iron and steel industries soon flourished in the city, drawing people to the valley in droves. The canal and railroads were instrumental to this growth, as noted by George Swank, editor of the *Johnstown Tribune*:

> *The whole character of the town suddenly changed. Canal boating and railroading took the place of flatboating; the Pennsylvania German element ceased to predominate in the makeup of the population; communication with other parts of the state and with other states became more frequent: homespun clothing was thenceforward not so generally worn: the town, at once, lost nearly all its pioneer characteristics.*[12]

By 1839, Johnstown was generating more income from the canal than any other town along the system. In tolls alone, the city collected $95,000 that year. By 1840, Johnstown and the surrounding area had a population of around three thousand. The community had a newspaper, drugstore, church and distillery; a handful of foundries and blast furnaces also operated in the city.[13]

By 1852, the Pennsylvania Railroad had laid tracks all the way from Philadelphia to Pittsburgh, and canal transportation was no longer necessary. The Pennsylvania Railroad connected to the Baltimore and Ohio Railroad system at various spots across the state, and commerce blossomed. The Cambria Iron Works paved the way for economic prosperity in Johnstown. Founded by George S. King in 1852—the same year that rail transportation through Johnstown was completed—Cambria Iron quickly established itself as one of the world's leading steel producers. By 1860, Johnstown had become the nation's leader in the industry, producing two-thirds of the country's steel. "In no part of the United States are found so many advantages for the manufacture of iron as at Johnstown," read a Cambria Iron flyer. "Millions of tons of iron can be made here without going three-quarters of a mile for any portion of the coal, ore and lime, or for the stone and brick for the furnace buildings and hearths."

The Western Division of the Pennsylvania Canal was a complex system of dams, locks, tow paths and aqueducts. *Johnstown Area Heritage Association.*

Cambria Iron relied heavily on its immigrant labor pool for all different kinds of jobs, including feeding its blast furnaces. *Johnstown Area Heritage Association.*

Immigrants who were looking for work poured into the city. They found jobs in Cambria Iron's mills, mines and coke-producing plants. Cambria Iron powered its mills with the bituminous coal that miners extracted from the vast deposits in the hills surrounding the city. Shortly before the Great Flood, Cambria Iron employed more than seven thousand men. Most of these laborers were immigrants who rented row houses from the company and earned around $1.50 a day. This labor pool, comprised primarily of German and Welsh immigrants, soon swelled the city's population to around thirty thousand.[14]

Slag from Cambria Iron's mills was dumped along the Conemaugh River to create more space for expansion. This narrowed the channel through which the river flowed. Smoke billowed from the company's stacks around the clock, lingering above the city and serving as proof of a thriving steel industry. The rich seams of coal in the mountains around Johnstown were heavily mined. These mountains were also stripped of their timber in many spots to provide Cambria Iron with lumber. The erosion of these mountains would contribute greatly to the pending disaster. This erosion, however,

Coal has been mined in the Johnstown area since the 1760s. The first large-scale mining in the region began in 1856 at the Rolling Mill Mine. *Johnstown Area Heritage Association.*

was of little concern to most Johnstowners in the years leading up to May 1889. For as long as business at the Cambria Iron Works boomed, there were plenty of jobs for folks throughout the region.[15]

Johnstown's residents were solidly working class. A handful of wealthy residents lived in the downtown section of the city, while coal miners and mill workers raised their families on the hillsides above. "Life in Johnstown meant a great deal of hard work for just about everybody," noted McCullough. "Not only because that was how life was then, but because people had the feeling they were getting somewhere. The country seemed hell-bent for a glorious new age, and Johnstown, clearly, was right up there, booming along with the best of them."[16]

THE CLUB ON THE HILL

Once rails were laid all the way from Pittsburgh to Philadelphia, the Pennsylvania Railroad abandoned the Main Line Canal and sold the South

Fork Dam, which had once been used in the system. This dam sat along the Conemaugh River, roughly fourteen miles upstream from Johnstown. The largest known man-made earthen dam in the world at the time, it held Lake Conemaugh. The dam had been built by state workers to provide a water supply for the canal route, and it was completed in 1853—fifteen years after the project began—at a cost of $190,000.

Engineer William Morris designed the dam, working closely with contractors James Moorhead and Hezekiah Packer. The earth and rock in the dam were ten feet thick at the top and more than 220 feet thick on the valley floor. During dry summer months, extra water from the lake was used to supplement the canal system. Locked in by steep mountain walls for about ten miles, the gorge between Lake Conemaugh and Johnstown is the nation's deepest river gap east of the Rocky Mountains. "The [transportation] solution seemed obvious enough," wrote McCullough. "Put a dam in the mountains, where it could hold a sufficient supply of water to keep the basin working and the canals open, even during those summers when creeks vanished, and only weeds grew." Following snowmelt each spring, when water tables reached their highest levels, Lake Conemaugh covered as much as 450 acres and reached depths of seventy feet. A number of streams and creeks poured into the man-made lake in addition to the south fork of the Little Conemaugh River, including Bottle Run, Muddy Run, Rorabaugh Creek, South Fork Creek, Toppers Run and Yellow Run. The South Fork Dam was seriously damaged on June 10, 1862, when a section of the stone culvert underneath it collapsed. Property upstream suffered only minor damage, but where the culvert collapsed, a large section of the dam was washed away. In 1879, cast iron valves and pipes from the dam were removed and sold as scrap.[17]

The South Fork Dam and five hundred acres of surrounding property were sold in 1879 to Benjamin Ruff, a Pittsburgh businessman who was representing a group of investors. A native of Schenectady, New York, Ruff had worked as a tunnel contractor with the Pennsylvania Railroad and had experience in real estate. His group's intent was to repair the dam and create a summer resort at the site. Led by Ruff and Pittsburgh's "king of coke," Henry Clay Frick, the group paid $2,500 for the dam and adjacent property. Ruff convinced his fellow investors—fifteen prominent businessmen from Pittsburgh—that this rural, mountainous site above Johnstown could be transformed into one of the finest summer resorts in the country. On October 15, 1879, work on the "South Fork Fishing and Hunting Club" began.

A crew of roughly fifty laborers worked under the direction of club members. These workers lowered the dam—which had been 72 feet high and 931 feet long—by 3 feet. They also widened it, converting the top of the dam into a road for horse-drawn carriages. They dug up and removed five iron runoff pipes that had been embedded in the dam's base. This created a 274-foot-deep hole, which was filled with mud, logs and rock. "The reconstruction of the dam," noted Peter Toner, an engineer who would inspect the dam years later, "was not remotely consonant with the elemental principles of engineering." A fishing screen was inserted into the spillway to contain one thousand stocked black bass in the man-made lake.[18]

Following these modifications, the dam stood just four feet above the spillway. Debris soon began collecting in the screens that kept the bass from escaping. Less than a mile upstream from the dam, on the western side of the lake, a forty-seven-room clubhouse was constructed, as well as cottages, boathouses and waterfront walkways. The sixteen "cottages" stood two to three stories tall, with ten to fifteen rooms each. The club's fleet of watercrafts included a pair of steam yachts. To say that the South Fork Fishing and Hunting Club stood in stark contrast to the community in the valley below would be a gross understatement. "It was a picture of life so removed from Johnstown that it seemed almost like a fantasy, ever so much farther away than fifteen miles, and wholly untouchable," wrote McCullough. "It was a picture that would live on for a long time after."[19]

On November 15, 1879, the club secured a charter, and annual membership fees were set at $800. South Fork was just a two-hour train ride from Pittsburgh, but the fresh mountain air of the Laurel Highlands made it seem like a world away from the grit and grime of the city. Almost all of the club's fifty or so members were connected in some way through Carnegie Steel. Andrew Carnegie, an international giant in the steel industry, owned a home roughly thirty miles northeast of Johnstown in the town of Cresson.

Carnegie was not only a giant in the steel industry but also a titan in the railroads, iron, oil and coal businesses. Born in 1835 in Dunfermline, Scotland, he and his family immigrated to the United States in 1848 to escape the poverty spreading across much of Britain and Scotland at the time. Upon arriving in western Pennsylvania, Carnegie's father, William, found employment in a Pittsburgh cotton mill. His son, Andrew, landed a job with the Pennsylvania Railroad at age twenty and quickly earned the respect of the company's president, J. Edgar Thompson. When the Civil War began in April 1861, Thompson assigned the young Carnegie to the

The South Fork Fishing and Hunting Club became a getaway not only for its members, but also for these businessmen's families. *Johnstown Area Heritage Association.*

role of superintendent of the Union's military railways. Following the war, Carnegie enjoyed a meteoric rise within the steel industry.[20]

Carnegie fell in love with the Laurel Highlands to the east of Pittsburgh and began spending large portions of each year at his Queen Anne–style summer home in Cresson. Named "Braemar," the house sat next to the Pennsylvania Railroad's massive Mountain House Hotel. For roughly a dozen years, Carnegie and his mother, Margaret, spent much of the months of June through October at their mountain retreat, often entertaining guests who stayed at the nearby hotel. On November 10, 1886, Carnegie's mother died at Braemar after contracting pneumonia. A grief-stricken Carnegie left the house in December 1886 and never returned. He remained fond of the Laurel Highlands, however, and soon turned his eye to nearby South Fork, becoming one of the new club's founding members.[21]

Soon after workers completed modifications to the South Fork Dam, it began springing leaks. These holes were patched with mud, straw and manure. Each spring, as water in the lake rose following snowmelt and mountain runoff, there was talk among Johnstowners that the dam might not hold. But it always had, and residents in the city below talked about this possibility casually—similar to everyday conversation about the weather.

Andrew Carnegie sold Braemar, as well as five hundred acres of land he had purchased in Cresson, after his mother died at the summer retreat in 1886. *Library of Congress, Prints and Photographs Division, Washington, D.C.*

The possibility of the dam breaking, however, was not a passing concern for the head of the Cambria Iron Works. Daniel Johnson Morrell commanded a presence in Johnstown from the moment he arrived in the city in 1855. Aside from leading Cambria Iron, the native of North Berwick, Maine, served as president of two banks and headed the city's water and gas companies. He presided over city meetings and served as president of the American Iron and Steel Association. Morrell was also active in politics, serving two terms in the U.S. Congress. He lived on Main Street in one of the city's largest homes and was described by McCullough as the "king of Johnstown."[22]

Morrell had left Maine in 1837 at the age of sixteen, following his older brother, David, to Philadelphia. David helped launch Trotter, Morrell and Company, a wholesale dry goods business, in the city. Daniel served as a clerk in his brother's business for five years before venturing out on his own. He became a partner at Martin, Morrell and Company, another dry goods enterprise in Philadelphia, one that specialized in high-end foods. Oliver Martin, his partner in this venture, died in 1854. A year later, Morrell headed across the state to Johnstown.

Under Morrell's leadership, the Cambria Iron Works grew rapidly. By 1878, the original one-acre site on the north bank of the Conemaugh River—about a half mile from the center of town—expanded to sixty acres, and the company's footprint across the region was massive. Cambria Iron owned 48,403 acres across seven counties, and it had laid sixty-eight miles of rails, either above or underground, in its mines. By 1880, the company employed roughly seven thousand men.[23]

Cambria Iron's rapid growth can be largely attributed to the company's perfection of the Bessemer Process, which, at the time, was the most inexpensive way to mass produce steel. In this process, air is blasted through molten iron, removing impurities through oxidation. The first commercial steel rails in the United States were produced at the Cambria Iron Works in 1867, and a massive Bessemer plant was completed in Johnstown in 1871.[24]

Cambria Iron also expanded its footprint beyond Pennsylvania, forming a partnership with a steel products company from Jersey City, New Jersey, owned by Josiah Gautier. In 1878, the Gautier Company established itself as a subsidiary of Cambria Iron and set up a base of operations on the south bank of the Little Conemaugh River. Three years later, Josiah Gautier dissolved the partnership, and Morrell and Cambria Iron took over this division, which specialized in the production of agricultural equipment.[25]

Morrell was staunchly antiunion, and Cambria Iron employees were banned from organizing. When national union leaders visited the valley and attempted to talk to Cambria Iron employees, Morrell ran them off. During his tenure as president of the company, his employees were among the lowest-paid steelworkers in Pennsylvania. Yet, despite his hard-nosed approach to dealing with his labor force, "King Morrell" was protective of his adopted city and its residents. When he heard about the modifications being made to the South Fork Dam, he became concerned. Morrell sent one of his staffers—engineer and geologist John Fulton—to inspect the changes that had been made. Following his inspection, Fulton sent a letter to Morrell:

> *There appear to be two serious elements of danger in this dam. First, the want of a discharge pipe to reduce or take the water out of the dam for needed repairs. Second, the unsubstantial method of repair, leaving a large leak, which appears to be cutting the new embankment.*
>
> *As the water cannot be lowered, the difficulty arises of reaching the source of the present destructive leaks. At present, there is forty feet of water in the dam, when the full head of sixty feet is reached, it appears to me*

Left: The "king of Johnstown," Daniel Morrell, had serious reservations about the modifications members of the South Fork Fishing and Hunting Club made to the South Fork Dam. *Johnstown Area Heritage Association.*

Right: John Fulton repeatedly warned Daniel Morrell that the South Fork Dam was a serious safety hazard following the modifications made to it by the club. *Johnstown Area Heritage Association.*

> *to be only a question of time until the former cutting is repeated. Should this break be made during a season of flood, it is evident that considerable damage would ensue along the line of the Conemaugh.*
>
> *It is impossible to estimate how disastrous this flood would be, as its force would depend on the size of the breach in the dam with proportional rapidity of discharge. The stability of the dam can only be assured by a thorough overhauling of the present lining on the upper slopes and the construction of an ample discharge pipe to reduce or remove the water to make necessary repairs.*

Fulton letter to Morrell
November 26, 1880[26]

Morrell forwarded these concerns to Benjamin Ruff, who dismissed the findings in a December 2 letter: "We consider his [Fulton's] conclusions as to our only safe course of no more value than his other assertions....You and your people are in no danger from our enterprise." Ruff's assurance did not satisfy Morrell. Three days before Christmas in 1880, Morrell returned correspondence:

> *I note your criticism of Mr. Fulton's former report and judge that in some of his statements, he may have been in error, but I think that his conclusions in the main were correct.*
>
> *We do not wish to put any obstruction in the way of you accomplishing your object in the reconstruction of this dam; but we must protest against the erection of a dam at that place that will be a perpetual menace to the lives and property of those residing in this upper valley of the Conemaugh from its insecure construction. In my judgment, there should have been provided some means by which the water would be let out of the dam in case of trouble, and I think you find it necessary to provide an outlet pipe or gate before any engineer could pronounce the job a safe one. If this dam could be securely reconstructed with a safe means of driving off the water in case any weakness manifests itself, I would regard the accomplishment of this work as a very desirable one, and if some arrangement could be made with your association by which the store of water in this reservoir could be used in time of drouth in the mountains, this company would be willing to cooperate with you in the work and would contribute liberally toward making the dam absolutely safe.*

Morrell letter to Ruff
December 22, 1880[27]

Morrell's concerns fell on deaf ears. Ruff was convinced that the people and property in the valley below the dam faced no serious threat, and none of Morrell's proposals were seriously considered. Frustrated and at a loss on how to spur any type of action, Morrell joined the club in order to keep tabs on what was going on in the mountain resort above his city. His membership in the club, however, was short-lived. Troubled by health issues, Morrell resigned from the Cambria Iron Works in January 1884, after serving as its president for twenty-nine years. On August 20, 1885—a little more than three and a half years before a deluge would destroy his adopted city—he died at his home in downtown Johnstown at the age of sixty-four.

As for John Fulton, he would not be the last engineer to visit the South Fork Dam. Nine years after his final inspection in 1880, teams of engineers from around the country would travel to the South Fork Fishing and Hunting Club seeking to determine what went so tragically wrong.

A LAKE UNLEASHED

The winter of 1888–89 hit Cambria County hard. In early April 1889, a foot and a half of snow fell on Johnstown. In early May, heavy rains fell frequently on the city, and flash flooding was a constant issue. On May 28, 1889, a storm formed over the Midwest and began moving eastward. When it settled above Johnstown two days later, it met up with two other storms and unleashed the heaviest downpour the city had ever recorded. Depending on the location, six to ten inches of rain fell over a twenty-four-hour period. The start of the storm mildly disrupted Johnstown's "Decoration Day" (now known as Memorial Day) festivities, which included a parade through the downtown section of the city during the late afternoon. "On Memorial Day 1889," noted Peter Toner, "the veterans of 1861 reverently paid tribute to their dead comrades, mercifully unaware that before another day had passed, many who lamented the dead would themselves be lamented."[28]

The skies cleared briefly before heavy rains again began falling around 9:00 p.m., continuing through the night. By the early morning hours of May 31, the Conemaugh River began surging over its banks. In a hillside farmhouse above the South Fork Dam, Elias Unger—who had succeeded Ruff as president of the club—looked out his window onto the swollen Conemaugh Lake around 6:00 a.m. The reservoir of water was close to cresting the dam, and Unger estimated that the water was about five feet from spilling over the top. He hurriedly assembled a team of men and told them to unclog the dam's spillway, which had become blocked by a broken fish trap, tree limbs, roots and other debris. When this failed, the men tried digging a spillway at the other end of the dam to relieve some of the water pressure, but this effort was also unsuccessful. Another crew of workers arrived and began working at the top of the dam, mounding dirt, mud and rocks to raise the height of the earthen wall. They also tried to pile mud and rock on the dam's face to save the eroding wall, but these ventures all failed.[29]

Unger's assembled workers abandoned their efforts to preserve the dam around 1:30 p.m. The club president ordered the men to move to higher ground and instructed a man named John Parke to travel by horseback to a telegraph tower in South Fork to send warning that the dam was going to fail. The club had telephone lines that connected it to South Fork, but they had been turned off for the winter and had not been reconnected. Telegraph lines in the region were already down, knocked out by the heavy winds that accompanied the storm. South Fork's telegraph operator, Emma Ehrenfeld, could only send Parke's message four miles down the valley, to the village of Mineral Point:

> *About noon, I judge it was, a man came in, very much excited. He says, "Notify Johnstown right away about the dam." He says, "It's raising very fast, and there's danger of the reservoir breaking."...We didn't have any wires then—our wires were all down, and I couldn't work with Johnstown direct.*
>
> *I called the operator at Mineral Point; he was the only one I could work with west. We fixed up a message, and I asked him to send it. He said he could send it west from there with one of the division men. I don't know how it was worded, but anyway, that there was danger at the reservoir. It was directed to the agent at Johnstown and the yard master at Conemaugh. He was to send it by a man to the next office west, and they were to forward it to Conemaugh and Johnstown by wire. Whether it ever reached Johnstown, I am unable to say.*[30]

The telegraph operator in Mineral Point, W.H. Pickerel, had a track man deliver the message a mile and a half down the valley. This message made it as far as East Conemaugh before Western Union wires stopped working around 3:00 p.m. At 3:10 p.m. on May 31, 1889, the South Fork Dam collapsed, unleashing twenty million tons of water. The water poured into the Little Conemaugh River with the force of Niagara Falls during the spring melting season. Conemaugh Lake was drained in forty-five minutes.

As it barreled toward Johnstown, the deluge picked up debris, including thirty-three locomotives, fifty miles of railroad tracks, and around 200,000 pounds of steel cables, as well as buildings, trees, houses, people and animals. At the Conemaugh Viaduct—a seventy-eight-foot-long sandstone railroad bridge—the water was stopped temporarily. More debris quickly accumulated, jamming against the bridge's arch. Against this weight, the viaduct collapsed, and the water continued down the valley. The water had gained renewed strength during the time it was bottled up at the viaduct, resulting in an even stronger force once it broke free.[31]

The raging waters hit the Cambria Iron Works facilities about a mile northeast of the center of Johnstown at a speed of around twenty-four miles per hour. In Old Conemaugh Borough, it swept up massive clusters of barbed wire, which became entangled with the other debris. Roughly a third of the Woodvale section of the city's 1,100 residents were killed when

The raging water wasted little time laying waste to various buildings at the Cambria Iron Works in Woodvale. *Johnstown Area Heritage Association.*

the water arrived. Boilers exploded, sending black smoke into the gray skies above the valley. Fifty-seven minutes after the South Fork Dam collapsed, the wall of water—which was traveling around forty miles per hour and reaching heights of sixty feet—slammed into downtown Johnstown.[32]

At the city's Stone Bridge, the debris in the floodwater formed a temporary dam, just as it had at the Conemaugh Viaduct, resulting in a surge that temporarily rolled upstream along the Stonycreek River. Soon, however, gravity prevailed, and the unleashed water proceeded on its path. A number of Johnstown residents became trapped in a fire at the Stone Bridge. The deluge had derailed several railroad cars carrying crude oil, and the oil saturated debris at the bridge. An inferno ensued. "Despite the water, which raged around it," wrote Toner, "the whole mass of debris soon became a roaring, crackling conflagration, forming a flaming breastwork for a dam of destruction and death. Death by fire in the midst of a flood." At least eighty people died in this fire, which burned for three days, and the smell of charred human flesh was overwhelming. When the floodwater receded, bodies and debris near the bridge covered thirty acres.[33]

The Stone Bridge was part of the Norfolk Southern Railway and was only a few years old at the time of the 1889 flood. It is estimated that 100,000 tons of debris collected at the bridge. *Johnstown Area Heritage Association.*

Around 6:00 p.m., Pennsylvania Railroad superintendent Robert Pitcairn telegraphed from Sang Hollow—a gap in the Laurel Highlands west of Johnstown—noting that he had seen around two hundred people floating by on gondola cars, shanties and other debris. Six-year-old Gertrude Quinn Slattery was one of those caught in the floodwaters:

> *I had great faith that I would not be abandoned....A large roof came floating toward me with about twenty people on it. I cried and called across the water to them to help me. This, of course, they could not do. The roof was big, and they were all holding on for dear life, feeling every minute that they would be tossed to death. While I watched, I kept praying, calling and begging someone to save me. Then I saw a man come to the edge, the others holding him and talking excitedly. I could see they were trying to restrain him, but he kept pulling to get away, which he finally did, and plunged into the swirling waters and disappeared.*
>
> *Then his head appeared, and I could see he was looking in my direction and I called, cried and begged him to come to me. He kept going down and coming up, sometimes lost to my sight entirely, only to come up next time much closer to my raft. The water was now between fifteen and twenty feet deep.*

The raging fire at the Stone Bridge burned for three days across thirty acres.
Today, the bridge is a landmark that is synonymous with Johnstown. *Johnstown
Area Heritage Association.*

*As I sat watching this man struggling in the water, my mind was firmly
fixed on the fact that he was my savior. At last, he reached me, drew himself
up and over the side of the mattress and lifted me up. I put both arms
around his neck and held on to him like grim death. Together, we went
downstream with the ebb and flow of the reflex to the accompaniment of
the crunching, grinding, gurgling, splashing, crying and moaning of many.
After drifting about, we saw a little white building standing at the edge of
the water, apparently where the hill began. At the window were two men
with poles helping to rescue people floating by. I was too far out for the poles,
so the men called: "Throw that baby over here to us!"...So, Maxwell
McAchren threw me across the water.*[34]

As the wall of water roared toward downtown Johnstown, Anna Fenn's
husband, John, was at a neighbor's house, helping him move furniture from
the family's first floor to its second floor. He was washed away when the
flood struck. In John and Anna's home, Anna and her seven children clung
to each other, but one by one, they were pulled apart. Only Anna survived.
"The water rose and floated us until our heads nearly touched the ceiling.

It was dark, and the house was tossing every way. The air was stifling, and I could not tell just the moment the rest of the children had to give up and drown." Anna gave birth to a baby girl a few weeks after the flood, but this child did not survive.[35]

Once the floodwater receded, everywhere one looked, there were scenes of death, loss and catastrophic damage. According to a reporter for the *Three Rivers Tribune*:

> *As the water recedes, a more thorough view of Johnstown's complete destruction is to be seen. The business part of the town is virtually wiped out of existence. Hardly a building of consequence escaped. The fine public library, one of the finest in the state, is no more. The grocery department of Wood, Morrell, and Co.'s store, the largest of its kind between Philadelphia and Pittsburgh, is gone. Jacob Swank's immense wholesale and retail hardware store is swept away. The new opera house is gone, and the unique skating rink was swept away also. The Hurlburt House, a large four-story structure, melted away like a pile of sand. The Mansion House and other large structures went the same way. Main Street is a total wreck. From South Fork to this spot in the heart of Johnstown, the way is swept almost clean.[36]*

Wood, Morrell and Company (*far right*) was the "company store" for the Cambria Iron Works. It was designed by Philadelphia architect Addison Hutton and was torn apart by the floodwater. The Cambria Iron Works General Office (*the building to the left*) was damaged but survived the flood. *Johnstown Area Heritage Association.*

Four square miles of the downtown area were completely destroyed after waves as high as forty feet hammered the city. *Johnstown Area Heritage Association.*

More than 1,600 homes were destroyed in the flood, and property damage was estimated at more than $17 million. *Johnstown Area Heritage Association.*

Along Clinton Street, every single structure was destroyed, with the exception of a section of St. John's Covenant Chapel. At the altar of this church—as floodwaters tore through the city beyond the chapel's walls—the Sisters of Charity knelt and recited their daily prayers.[37]

THE AFTERMATH

The 1889 Johnstown flood was the worst flood to hit the United States in the nineteenth century. It killed 2,209 people, and roughly one-third of the dead were never identified. Their remains were buried in the "Plot of the Unknown" in what is now Grandview Cemetery in Westmont, a suburb above Johnstown. This cemetery had been purchased in 1885 by the Citizens Cemetery Association, a group comprised of officials at the Cambria Iron Works. It had only been open for a year before the Great Flood.[38]

It was not until July 10, 1889, that a day passed without the recovery of a body. Sixteen people in Mineral Point were killed, and more than three hundred died in Woodvale. Most of those who died, however, lived or worked in the bottom of the valley, where the water had nowhere to go. Downtown churches and schools became temporary morgues. In an effort to estimate the number of victims, survivors were asked to register at various locations in the days after the water subsided.[39]

General Daniel Hastings led the post-flood relief effort. A native of Bellefonte, Pennsylvania, he later served as governor of the state from 1895 to 1899. Hastings decided he needed armed troops to maintain law and order. He asked Pennsylvania governor James Beaver for help. With Beaver's blessing, Hastings established a state of martial law in the city and its surrounding boroughs. Because of the massive piles of mud and debris that had collected, most of the city's cemeteries were inaccessible. Many of the victims were buried in shallow graves in Prospect, a suburb a half mile north of the downtown section of the city. Most of those interred in Prospect were relocated to Grandview Cemetery months later. The flood caused an estimated $17 million in damages, which equates to $486 million in 2021. Four square miles of the city's downtown and more than 1,600 homes were destroyed. Amazingly, however, several of the city's most prominent businesses quickly reopened. The Cambria Iron Works resumed partial operations on June 6, 1889. It returned to full operation within a year and a half. By July 1, a number of storefronts on Main Street had also reopened.[40]

The bodies of more than 750 victims of the 1889 flood were never identified. Many of those who survived had to pick up the pieces of their lives without their loved ones. *Johnstown Area Heritage Association.*

Where and how to start cleaning up once the floodwater receded was difficult for survivors to determine. In many cases, all of their earthly possessions were gone. *Johnstown Area Heritage Association.*

Five years after leading the clean-up efforts following the Great Flood, Daniel Hastings was elected governor of Pennsylvania. *Johnstown Area Heritage Association.*

Support in the form of manpower, food, medical supplies and financial aid poured into the valley almost immediately after the disaster. Around 10:00 p.m. on May 31—the day of the flood—a train carrying volunteers arrived in Sang Hollow. In the weeks that followed, every state sent some type of relief aid, and eighteen countries assisted Johnstowners in some way. Inmates at the Western Penitentiary in Pittsburgh baked one thousand loaves of bread a day and shipped them to Johnstown. The Standard Oil Company donated kerosene to survivors. Cincinnati grocers sent twenty thousand pounds of ham. An estimated 1,400 railroad cars of goods—weighing 17 million pounds—made their way to the city during the second half of 1889. The American Red Cross was also an important part of the recovery effort. Clara Barton, a nurse who served on various battlefields during the Civil War, had founded this international relief agency in 1881 and she would lead it for the next twenty-three years. Barton and fifty other

Above: Nine years
after the 1889 flood,
the Cambria Iron
Works reorganized
as the Cambria
Steel Company.
*Johnstown Area
Heritage Association.*

Right: Clara Barton
did not leave
Johnstown until
October 24, 1889,
when she felt that
the Red Cross
had completed
its mission in the
city. *Johnstown Area
Heritage Association.*

Red Cross volunteers arrived in Johnstown on June 5 and stayed in the valley for the next five months. They established six Red Cross "hotels" throughout the city.[41]

Volunteers were not the only ones to arrive in the valley. More than one hundred newspapers and magazines from around the country sent correspondents to the city. These reporters focused their coverage—and their wrath—on members of the South Fork Fishing and Hunting Club. The *Chicago Herald* ran an editorial cartoon showing members of the club drinking champagne on a porch of the clubhouse while the flood was flattening Johnstown. A *New York Times* headline read, "An Engineering Crime—The Dam of Inferior Construction, According to the Experts." A June 7 *New York World* headline told readers that "The Club Is Guilty." Richard Burkert, president and chief executive officer of the Johnstown Area Heritage Association, captured the sentiment of many following the 1889 flood: "This was when the term 'robber barons' was coined. The concern at the time was: can you have a viable democracy when great wealth controls the political process? And even in the case of a horrible accident like this, you have a group of people who are above the law, which is the way Victorian Americans saw this."[42]

Despite widespread public consensus that blame for the dam's break belonged squarely on the members of the South Fork Fishing and Hunting Club, none of these men were ever held legally responsible for the disaster. Andrew Carnegie was in Paris when the dam broke. Upon hearing the news, he announced he would send $10,000 to the city to aid in recovery. Thirty-five of the sixty members of the club contributed financially to relief efforts. Regarding any culpability in the dam's break, however, they remained silent.[43]

Multiple lawsuits were filed against the club's members, but none were successful, and no compensation was granted to any of the disaster's survivors. Various courts ruled that the dam's break was an "act of God." Nonetheless, the press continued to hammer these men in the court of public opinion. They were vilified for not only their perceived responsibility in the disaster but also for designing a financial structure that kept their personal assets separate from club business in order to avoid financial liability.

In the first issue of Johnstown's *Tribune* newspaper following the disaster, editor George Swank wrote that "we think we know what struck us, and it was not the work of Providence. Our misery is the work of man." Most of the citizens of Johnstown and others around the country who followed the story agreed. Gertrude Quinn Slattery wrote the first book on the tragedy. It is titled *Johnstown and Its Flood* and was published in 1936:

Morgue workers traveled to the city to help embalm and lay to rest the more than two thousand victims of the Great Flood. *Johnstown Area Heritage Association.*

A group of men stand in the breach of the South Fork Dam after it yielded to the wall of water it once contained. *Johnstown Area Heritage Association.*

A colony of railroad and steel company officials protected the birds and other forest folk and stocked the dam with a variety of game fish. The fish disported themselves in these cool waters and they grew and multiplied and became important, so important in fact, that when the days and days of heavy rain drenched the hills, turned the rivulets into raging torrents, and sent them hurtling down to pour their fullness into the reservoir, the gentle keeper felt he could not open the flood gates, lest the valuable fish escape.[44]

Investigators ultimately concluded that the original designers of the South Fork Dam—all of whom were dead by 1889—had designed a reservoir without regard to "freak storms." There was little regard paid to other factors that contributed to the catastrophe. Few had blinked an eye as workers stripped the mountains above the city of their trees in order to provide lumber for Cambria Iron's building projects. This lumbering robbed the hillsides of much of their topsoil, soil that had absorbed snowmelt and water during the spring rainy seasons. Without enough topsoil performing this function, more and more water collected in the South Fork Dam. Cambria Iron's dumping of slag and other types of landfill off the banks of the Stonycreek, Little Conemaugh and Conemaugh Rivers was also a contributing factor in the tragedy. With narrower channels than Mother Nature had intended, water poured over these rivers' banks in May 1889 much sooner than it would have years earlier under similar conditions.[45]

The South Fork Fishing and Hunting Club reopened for business in the summer of 1889, but there was little interest among its members and their families in returning to the mountain retreat following the disaster. Club membership broke up entirely in 1904, and an auction sale was conducted shortly thereafter. Frank Shomo, the last known survivor of the 1889 flood, died in March 1997 in Blacklick, Pennsylvania, at the age of 108. Shomo was one hundred days old when the dam broke.[46]

2
JOHNSTOWN'S SECOND ACT

J ohnstown entered the twentieth century rebuilt, by and large, following the Great Flood, and the recovery was evidenced by many measures. The city's population grew from 21,805 in 1890 to 35,936 in 1900, and it would increase to more than 55,000 by 1910. It ranked as the ninth largest city in Pennsylvania by 1920, with a population of 67,327 residents.[47]

This growth can be attributed, at least in part, to consolidation. In November 1889, just months after the Great Flood, Johnstown consolidated with the communities of Millville, Cambria City, Prospect, Woodvale, Grubbtown, Moxham and Conemaugh to become a larger single city. Shortly thereafter, the leaders of Walnut Grove, Roxbury and Rosedale, as well as parts of Stonycreek and Upper Yoder Townships, asked for their communities to be annexed to the city, and they were added. In February 1890, residents of the expanded city elected Johnstown's first mayor, W. Horace Rose.[48]

The Cambria Iron Works continued to serve as an economic engine for the city and surrounding region. The company reorganized under the name Cambria Steel in August 1898, after flirting with the idea of moving its base of operations to a site in the Great Lakes region. Once the decision to stay in Johnstown was made, company leaders built a state-of-the-art mill in the Franklin Borough east of the city, and a wire plant was erected in Johnstown's Morrellville neighborhood shortly thereafter.

Over the first two decades of the new century, Cambria Steel spent $70 million on construction and improvements to its existing facilities

Johnstown's population increased to more than 30,000 by 1890, in large part due to the jobs Cambria Steel provided to area residents. *Johnstown Area Heritage Association.*

and equipment. By 1900, Johnstown ranked third in the nation in steel production, and its mills employed more than 16,500 people. Rails produced in Johnstown were shipped across the country and around the world. In Russia, tracks produced in Johnstown were used in the construction of the Trans-Siberian Railroad System. In 1910, Cambria Steel began construction of a Rod and Wire Division on a twenty-one-acre site across the Conemaugh River from the Coopersdale section of the city, right next to tracks that belonged to the Pennsylvania Railroad.[49]

KING COAL

Where there are mills, there are mines, and Johnstown was no exception. As noted by author John Strohmeyer in his book *Crisis in Bethlehem*, the Johnstown region was especially well-suited for both: "Nature provided the tremendous quantities of water necessary for steel operations and bountiful supplies of bituminous coal could be mined from deep veins practically underneath the plant." Coal had been mined in Cambria County since the 1830s, and large-scale mining in the valley began when the Cambria Iron

Works opened the Rolling Mill Mine in 1856. Producing coal during this period required grueling labor. Most miners worked at least ten-hour shifts, six days a week, and were paid by the tonnage of coal they extracted from the mines. These "pickminers" spent much of their workdays lying on their backs, attempting to loosen coal from seams above their heads. Like many of Johnstown's mill workers, the laborers in these mines were mostly of Eastern European descent.[50]

The first coal from the Allegheny Mountains of western Pennsylvania was extracted in 1768, after several businessmen purchased land near Pittsburgh from a group of Natives. Less than ten years later, coal from these underground seams was used to forge weapons for George Washington's Continental army. Following the Revolutionary War, the need for domestic coal spiked during the War of 1812 due to an embargo on shipping, which severely limited the amount of overseas coal arriving in the United States. With this increased demand came other challenges; foremost among them was the issue of transporting "black gold" to where it was needed.[51]

The owners of many of the mines that sprung up in western Pennsylvania during this period could not efficiently move their coal more than a few miles because there was no transportation infrastructure. The roads in the region had not been built to bear the weight of coal. It was at this point that business and government leaders began to consider a canal transportation system. By 1820, a workable system had been established across Pennsylvania and Maryland.[52]

With the canals providing reliable transportation, the coal industry in western Pennsylvania blossomed. When the first official statistics on coal production were compiled in 1838, Pennsylvania's secretary of the commonwealth reported that Pittsburgh companies had contributed $565,200 to the city's economy. Three years later, in 1841, coal mining was a million-dollar industry in Pennsylvania and Maryland. By the end of that same year, railroad tracks had been extended across the entire state, and the use of the canal system dramatically declined before disappearing altogether. Coal company executives became very wealthy over the next several decades, and the mining towns that they established were planned with detailed precision.[53]

When the South Fork Dam broke in May 1889, around two dozen mines in the Johnstown region were producing roughly 1 million tons of coal a year. Across all of Cambria County and neighboring Somerset County to the south, more than 100 mines were in operation at the time of the Great Flood. These mines produced more than 10 million tons of coal a year and

Above: Pennsylvania remains one of the leading coal-producing states in the United States, more than 250 years since coal was first extracted from a seam near Pittsburgh in 1768. *Johnstown Area Heritage Association.*

Opposite: The Rolling Mill Mine was a drift portal mine, or an underground mine in which the entry point is on the slope of a hill. *Johnstown Area Heritage Association.*

employed around fifteen thousand people. Following the flood, growth in the mining industry in the mountains around the city exploded. From 1900 to 1925, around 130 mines in the region produced more than 16 million tons of coal a year. The Great Depression slowed the coal industry's growth during the 1930s, however, and many of the region's smaller mines—and even some of its larger ones—closed. The Rolling Mill Mine closed in 1931, after more than 22 million tons of coal had been extracted from its seams. At its peak, this mine produced around 3,000 tons of coal a day.[54]

The Rolling Mill Mine was the site of one of the deadliest mining accidents in U.S. history. On July 10, 1902, an underground explosion in the mine killed 114 men. The explosion was attributed to a firedamp—a methane gas mixture that collects in air pockets in many bituminous coal mines. "No list of the names of the dead miners can be given," said Powell

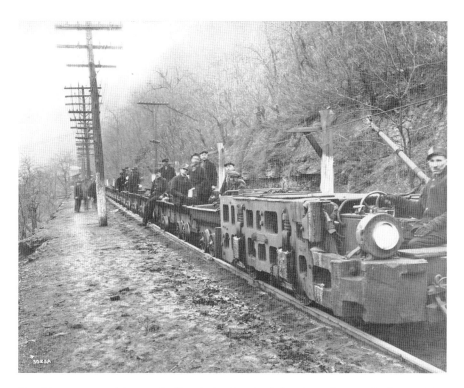

Stackhouse, president of Cambria Steel, following the explosion. "The majority of the miners were foreigners and were known only by check and not by name. The only way their names will ever be known—if the bodies are recovered in time for identification—will be by their families sending their names to us." A check was a metal tack pinned to miners' shirts. It was used for identification following accidents.[55]

CLASSED CITY

Cambria Steel's increasing need for laborers in its mills and mines throughout the early twentieth century was met primarily by immigrants from eastern and southern Europe. From 1890 to 1910, the number of immigrants from these parts of Europe who settled in the Johnstown area increased from 2,400 to more than 12,000. Like other communities across the United States, a hierarchy emerged among the immigrants who settled in Johnstown. The Germans, Irish, Scots and Welsh ranked highest on the social and employment ladders, with Bohemians, Croatians, Hungarians, Italians, Poles, Serbians, Slovaks and Slovenians closer to the bottom.[56]

Many of these immigrants settled in Cambria City, Minersville and other working-class neighborhoods near Johnstown. These newcomers faced frequent discrimination. George Swank, editor of the *Tribune*, expressed the sentiment of many of the city's residents in a 1908 article: "In the last thirty or thirty-five years, the lack of homogeneity among the people of Pennsylvania has been conspicuously and most painfully emphasized in the invasion of large sections of the state by hordes of Italians, Slavonians and other immigrants of distinctly lower types." While Johnstown was becoming much more ethnically diverse following the turn of the century, the city's racial diversity lagged behind many other industrial cities across the United States. According to 1900 census data, 314 Black citizens lived in the city at the turn of the century, representing less than 1 percent of the total population. Across all of Cambria County in 1900, less than one half of 1 percent of residents were Black.[57]

When World War I began in 1914, European immigration to the United States slowed due to federal restrictions. The mills, mines and factories of Johnstown and other industrial cities, however, needed more laborers. Many of these cities turned to Black people from the South to meet their labor needs. Cambria Steel sent recruiters south in search of workers. By 1920, Black citizens made up 10.9 percent of the labor force in steel mills across Pennsylvania.[58]

Johnstown's Black population increased from roughly 300 in 1900 to 1,650 in 1920. In many instances, these new members of the community were not met with open arms. In a 1903 article in the *Democrat*, reporter Nelson Raynor quoted a Black Johnstowner who described discrimination in the city: "The colored man must remember his position and must not think he is better than the white man….He has a place of his own, so long as he stays in his place and attends to his business."[59]

Ku Klux Klan activity in the area increased with the arrival of Black laborers from the South. In January 1922, a story in the *Tribune* noted that "a large class of prominent men" had been initiated into the organization at a cross burning. By 1925, Johnstown Klavern No. 89 had approximately 1,800 members. Throughout the 1920s, Johnstown KKK leaders hosted "Klan Days" in the city, bringing as many as 30,000 Klansmen and their families to the valley. Johnstown was also one of the first cities in Pennsylvania to establish Klavaliers, the military "secret police" units of the Klan.[60]

Kelly Miller, a professor of sociology at Howard University in Washington, D.C., visited Johnstown in the spring of 1923 to explore the living conditions of the city's Black residents:

I have visited all types and kinds of communities in which Negroes live in all parts of the United States....I have seen them in alleys and shady places; I have witnessed their poverty and distress in city and country. But I can truthfully say that it has never been my good fortune, or misfortune, to look upon such pitiable conditions as prevailed in Johnstown.[61]

STRIKE OF 1919

In February 1916, Philadelphia's Midvale Steel and Ordnance Company—one that boasted a national reputation for producing high-quality steel—purchased Cambria Steel. Midvale soon began manufacturing freight car wheels in Johnstown. The company poured more than $100 million into upgrades in the city, ramping up operations on many different fronts. Midvale quickly became a national leader in the production of coke, steel freight cars, wheels, axles, gears, flywheels, forgings, bars and all different kinds of wire products. The company expanded its facilities in Johnstown, widening its footprint nine miles up and down the Conemaugh and Little Conemaugh Rivers. Under Midvale president William Corey's direction, three new blast furnaces were built in the city, and production at the plant boomed throughout World War I. When the war ended in 1918, however, operations in Johnstown—and in other factory cities around the country—slowed significantly. In 1917, Midvale's profits totaled $35.6 million across its holdings. Just four years later, it operated at a loss of $5.3 million.[62]

In January 1919, leaders of the National Committee to Organize Iron and Steel Workers began visiting steel-producing cities across the Northeast, Mid-Atlantic and Midwest to recruit new members. This effort was led by Thomas Conboy, William Foster and Samuel Gompers. Mine and mill workers around the country had grown increasingly frustrated with low wages, poor living conditions, long shifts and the lack of any real collective bargaining power.[63]

On April 1, 1919, a crowd of more than 3,500 marched in a "labor parade" in Johnstown—one organized by Conboy. Three months later, members of the national committee voted on whether or not to strike, and 98 percent of the group elected to walk off the job. President Woodrow Wilson, as well as Gompers, wanted to postpone a national strike until the various stakeholders were better organized, but Foster believed the time to walk off the job had arrived—and he held enough clout to make it happen. On September 22, 1919, roughly 300,000 mill workers and miners around the nation refused to report to work.

In Johnstown, most of the area's mine and mill workers honored the strike call, and fifteen thousand Midvale employees walked off the job. Midvale officials refused to negotiate with the striking workers, and both sides dug in their heels. An Italian immigrant named Dominick Gelotte from the mining village of Nanty-Glo just north of Johnstown was an important figure in the labor movement in the area. Gelotte shared the following message in a flyer to Midvale strikers:

> *Brothers: You certainly realize the hard struggle now going on in Johnstown, Pa. between the Cambria Steel* [Midvale had retained the name "Cambria Steel" for its Johnstown operations] *and its employees. For sixty-nine years, the workers of the said city have been under the wings of this tyrannical corporation, and no workingman was free but a secular slave of this steel corporation. For the first time in the history of Johnstown, the workers have taken a determined position to fight for their exclusive rights.*[64]

The strikers also received support for their cause from fellow laborers who did not work for Midvale. Two thousand coal miners who worked for a subsidiary company of Midvale walked off the job, demanding union recognition as well. Laborers for the Pennsylvania and Baltimore and Ohio Railroads refused to haul materials and finished products into or out of Midvale's plants in a show of solidarity with the company's striking mill workers and miners.[65]

Despite this support, by November 1919, management had secured the upper hand in the standoff, and Johnstown's striking workers were on the defensive in the court of public opinion. A "citizens' committee" had been organized, and this group—which included prominent business leaders and clergy members from around the city—argued that the striking laborers should report back to work.

Midvale officials announced that the company would welcome back striking employees if they renounced their union affiliations. The company ran full-page advertisements in Johnstown's newspapers, urging workers to return to their jobs. A local group of union leaders that called itself the "Johnstown Strike Executive Committee" responded to Midvale's advertisements by running advertisements of their own, harshly criticizing working conditions at the company. These advertisements asserted that the workers had a right to unionize, but they also noted that strike leaders were willing to meet with Midvale officials. Union leaders urged Johnstown's citizens' committee to stop meddling in the dispute.[66]

John Brophy (*center*) emigrated with his family from England to the United States in 1892. Before he turned twelve, Brophy began working alongside his father in Pennsylvania coal mines. *Library of Congress, Prints and Photographs Division, Washington, D.C., photograph by Harris & Ewing.*

William Foster visited Johnstown for a rally scheduled for November 7. When he arrived in the city on the preceding evening, two newspaper reporters warned him to leave immediately, fearing that violence would erupt because of his presence. As Foster approached Johnstown's city hall, members of the citizens' committee intercepted him and escorted him onto a railroad car that carried him out of the valley.[67]

Foster's experience did not deter other strike organizers from visiting Johnstown. National labor leader Frank Kurowski visited the city and tried to get the U.S. Labor Department to mediate the standoff between Midvale and the striking workers, but the citizens' committee forced him out of town as well. After Kurowski's experience, John Brophy—president of United Mine Workers District 2—sent a telegram to Pennsylvania governor William Sproul. Brophy told him that the Johnstown Chamber of Commerce was denying miners their freedom of speech. The union leader also sent a letter to President Woodrow Wilson, requesting his help in arranging a meeting between the striking laborers and members of Midvale's management team.[68]

With both sides digging in their heels, and the strike dragging into a second month, Governor Sproul took action, sending state police officers to Johnstown. The officers' instructions were to maintain law and order. Several of these state troopers were given orders to protect the homes of Midvale executives. Harvey Thomas, president of Johnstown's General Strike Committee, accused troopers of rushing into a crowd of peaceful strikers in Cambria City on November 15. Thomas also claimed that troopers had invaded labor leaders' homes without cause and arrested striking workers on trumped-up charges.[69]

The press in Johnstown stood solidly in support of management throughout the strike, with stories in the *Tribune* and *Democrat* newspapers striking a distinctly anti-union tone. Most of the religious leaders in Johnstown also sided with Midvale officials. Clergy members described the labor leaders who visited Johnstown as radical and un-American outsiders who stirred up trouble and then departed without having to deal with the consequences. Reverend Eugene Garvey, bishop of the Altoona-Johnstown Roman-Catholic Diocese, criticized strike organizers and told the striking mill workers and coal miners who belonged to his diocese that they should work harder at saving their money, rather than demanding higher wages. An exception among the clergy officials who lined up in support of Midvale's management was Reverend George Dono Brooks of Johnstown's First Baptist Church. Brooks told his congregants that he supported the strikers' efforts. The pastor criticized the "social evils" that he believed were the root cause of the strike. After he publicly expressed his support for the striking steelworkers, elders in Brooks's congregation dismissed him from his pastoral responsibilities.[70]

On November 17, 1919, most of Midvale's mines and mills reopened. Striking employees, desperate for money after two months without a paycheck, began heading back to work. The strikers' solidarity unraveled as their bank accounts depleted. They were also growing increasingly weary of the attacks on their collective character by members of their communities. Nationally, the first workers to return to the mills and mines were in Chicago. Miners in Johnstown; Wheeling, West Virginia; and Youngstown, Ohio, soon followed. Striking workers in Cleveland and Pittsburgh held out the longest, but by December, most of the striking laborers in these cities had returned to their jobs as well. In the end, the only

Mill workers and miners in Pittsburgh were among the last laborers to remain on the picket lines during the national strike of 1919. The strike effort collapsed in early January 1920. *Library of Congress, Prints and Photographs Division, Washington, D.C.*

concession that Midvale officials made to their labor force in Johnstown was the institution of an eight-hour work shift.[71]

Three years later, in November 1922, Midvale sold its holdings in Johnstown to the Bethlehem Steel Corporation. The boom period that World War I had generated for the company had subsided by 1922, and Midvale's profits plummeted. The company sold off all of its holdings, with the exception of its original location in Philadelphia. By this time, Johnstown had become one of the largest industrial cities in the Mid-Atlantic region, boasting a population of more than 105,000. Bethlehem's parent company, Saucon Iron, had been founded in the eastern Pennsylvania city of Bethlehem in 1857. "When our company took over the holdings of Cambria Iron and Cambria Steel, with it came a lot of real estate," said Charles Weidner, who worked in the real estate division of Bethlehem for more than four decades. "The mountainsides [above Johnstown] were covered with timber, which you can use. You cut the timber to put in the mines to hold the ceiling up. There were coal mine areas, timber areas, even natural gas wells and a lot of houses in the city."[72]

Left to right: Bethlehem Steel lawyer Paul Cravath with Charles Schwab and Eugene Grace. *Library of Congress, Prints and Photographs Division, Washington, D.C.*

The sale of Midvale's holdings to Bethlehem Steel was finalized on March 30, 1923, in a meeting in New York. "Bethlehem, Midvale and Cambria interests, today, completed all matters necessary for the consummation of the purchase by Bethlehem of the properties and assets of Midvale and Cambria," read a press release announcing the transaction. Shortly after the deal was finalized, Bethlehem Steel president Eugene Grace visited Johnstown. He noted that former Bethlehem Steel president Charles Schwab, who owned a summer estate in the community of Loretto, located thirty miles north of Johnstown, would take an "active interest" in operations in the valley.[73]

CHARLES SCHWAB

Charles Schwab had deep roots in Cambria County. He was born on February 18, 1862, in the town of Williamsburg in Blair County, the son of John and Pauline (Farabaugh) Schwab. The Schwab family relocated to neighboring Cambria County when Charlie was twelve, settling in Loretto. John operated a livery stable in this village and delivered the mail. The oldest of John and Pauline's eight children, Charlie helped his father with his mail carrier responsibilities. After completing grade school, Charlie pursued high school courses at Saint Francis College, a Roman-Catholic school near the family's home in Loretto. In 1879, he moved to the Pittsburgh area, where he landed employment at a grocery and dry goods store in Braddock.[74]

Braddock is about ten miles from Pittsburgh and served as the home of Edgar Thompson Steel Works, which was owned by Andrew Carnegie. William Jones served as Carnegie's general superintendent at the plant and was a regular shopper at the grocery store where Charlie worked. Jones struck up a friendship with the personable Schwab. He offered Schwab a job on a surveying crew, and Charlie accepted, quickly demonstrating his work ethic and ambition. Schwab was soon promoted to draftsman and then to engineer. Jones also assigned him the responsibility of providing a daily report on plant operations to Carnegie.[75]

Like Jones, Carnegie was drawn to Schwab's infectious personality, unbridled confidence and charm. The steel tycoon offered the twenty-four-year-old Charlie Schwab the general superintendent job at his Homestead Works plant. When Jones was killed in a furnace explosion three years later, Carnegie appointed Schwab head of the Edgar Thompson plant. At twenty-seven years old, Schwab was placed in charge of day-to-day operations at the largest steel plant in the United States.[76]

Schwab's power of persuasion carried him a long way in his career. He once bragged that he had borrowed more money on less collateral than anyone he knew. "I once heard a story which shows how strong his influence was," said Gertrude Fox, who worked as an industrial biologist and inspector at Bethlehem Steel from 1945 to 1947.

He needed a large amount of money, so he went to New York to a bank to borrow it. I'll make up the amount—$3 million. He spoke to the president of the bank and convinced him that he needed it for steel company purposes. He was so persuasive that the president said he could have it. The next day, he got a call from the president of the bank, saying, "You've got to come

Charles Schwab had a keen business sense that impressed Andrew Carnegie immediately. He told Carnegie that "the way to get things done is to stimulate competition." *Library of Congress, Prints and Photographs Division, Washington, D.C.*

back and talk to our board of trustees. When I told them the story that you told me, about what you were going to do with the money, it didn't come out the same."[77]

Schwab and Carnegie's relationship grew close as the two men dealt with one challenge after another, including labor strikes, accusations of defrauding the federal government and increased competition from other American steel companies. In 1897, Carnegie named Schwab the president of Carnegie Steel. In 1901, Schwab brokered a merger between Carnegie Steel and the steel holdings of J. Pierpont Morgan. The U.S. Steel Corporation was born out of the merger, and thirty-nine-year-old Charlie Schwab served as U.S. Steel's first president.[78]

Schwab purchased the Bethlehem Steel Company with his own personal fortune in May 1901, six weeks before the merger that formed U.S. Steel. At the time, Bethlehem employed around four thousand men and produced steel

Two furnaces continue to operate today at the Edgar Thompson Steel Works. *Library of Congress, Prints and Photographs Division, Washington, D.C.*

forgings for guns and marine engines, as well as armor plates. The plant sat along a mile-and-a-half stretch of the Lehigh River in eastern Pennsylvania. When J.P. Morgan learned that Schwab had purchased Bethlehem Steel, he was not pleased. Perhaps believing that divided allegiances would compromise Schwab's ability to lead U.S. Steel, Morgan bought Bethlehem from Schwab for $7.2 million, roughly the same amount that Schwab had paid for the company. Morgan owned Bethlehem Steel for roughly a year before selling it back to Schwab, who promptly sold it to a company called U.S. Shipbuilding.[79]

The next two years did not prove profitable for U.S. Shipbuilding. On June 30, 1903, a federal judge ruled that the company was insolvent and accused Schwab of "ruinous extortion." On August 4, 1903, Schwab resigned from his position as president of U.S. Steel. Sixteen months later, in December 1904, the Bethlehem Steel Company became the Bethlehem Steel Corporation, with Schwab as its president. "I intend to make Bethlehem the prize steelworks of its class, not only in the United States but in the entire world," said Schwab. "In some respects, the Bethlehem Steel Corporation already holds first place. Its armor-plate and ordnance shops are unsurpassed, its forging plant is nowhere excelled, and its machine shop is equal to anything of its kind."[80]

Schwab was a forward-thinker, a risk-taker and an innovator, and he brought these traits to Bethlehem Steel. According to John Strohmeyer, author of *Crisis in Bethlehem*:

> *Schwab built Bethlehem into a strong company largely because he was the only steel producer who had the courage to mortgage his future for a new type of mill to make structural beams—wide flange shapes that were stronger and cheaper than the old riveted girders. The towering skylines of many American cities are, today, a testimony to Schwab's vision. The other steel companies adopted innovation only after he proved that it worked.*[81]

Schwab's risk-taking and innovation translated into large profits for Bethlehem Steel. When he bought the company, it produced less than 1 percent of the nation's steel; by World War II, Bethlehem was second only to U.S. Steel in domestic steel production. The company established major hubs in Bethlehem and Johnstown, as well as Steelton, Pennsylvania; Buffalo, New York; and Sparrows Point, Maryland.[82]

Schwab enjoyed basking in his wealth and fame as much as he enjoyed the wheeling and dealing of the steel industry. His mansion on Riverside Drive in Manhattan was the largest private residence in the Big Apple. The $8 million home he and his wife, Rana, shared included ninety bedrooms, an art gallery and a power plant. In 1919, work on a second estate in Schwab's hometown of Loretto was completed. This estate was named "Immergrun"—German for evergreen—and it sat on one thousand acres of rolling hills. It included a golf course and seventeen other buildings.[83]

When business was booming for Schwab and Bethlehem, it really boomed. But there were also bust periods for the company under his watch. By 1936—the year of Johnstown's St. Patrick's Day flood—Schwab and Bethlehem were trying to weather a severe downturn in the industry. The first two decades of the twentieth century had been incredibly profitable for Bethlehem, but when the stock market crashed in 1929, its tremors were felt across the company's holdings. Bethlehem produced 55 million tons of steel in 1929; two years later, the company's steel output was 40 million tons. By 1932, the bottom had fallen out for Bethlehem and the national steel industry: 90 percent of the roughly 500,000 steelworkers across the United States were either working part time or were out of work altogether. Despite possessing the capacity to produce half of the world's steel, U.S. companies accounted for just 27 percent of global steel sales in 1932. Cheaper, imported

Andrew Carnegie owned a mansion on New York City's upper Fifth Avenue, but it paled in comparison to Schwab's Riverside Estate (*above*). "Have you seen that place of Charlie's?" Carnegie once asked a friend. "It makes mine look like a shack." *Library of Congress, Prints and Photographs Division, Washington, D.C.*

Today, Schwab's Immergrun Estate serves as a home to practicing and retired Franciscan priests and brothers. The estate is next to Saint Francis University, one of the oldest Catholic schools of higher education in the United States. *Saint Francis University Library.*

foreign steel created serious financial problems for American steel companies during the Great Depression.[84]

Schwab and Eugene Grace cut the wages of Bethlehem workers by 10 percent in 1931 and 15 percent in 1932 while pleading with the federal government to raise tariffs on imported steel. Yet neither man made any adjustment to his own annual salary. Schwab continued to collect $250,000 a year, and Grace's fixed annual salary of $180,000 did not change. The two men blamed the government for the company's economic woes: "We have done our part," said Schwab. "We have put our house in order. The Federal Reserve is doing its part. But above all, the federal government, which is the heart of our national structure, must balance its budget and restore confidence there."[85]

While the downturn in domestic steel production during the Great Depression was certainly consequential, the most significant event that impacted Bethlehem Steel during the 1930s was the rise of the U.S. labor movement. Organized labor had made strides from 1910 to 1920, but the movement stalled during the 1920s. In 1930, only 3.4 million workers in the United States belonged to a trade union. That number was about to grow exponentially, and for companies that had long sought to keep their workers from organizing—such as Bethlehem Steel—trouble loomed. While another boom period for Bethlehem was around the corner (World War II), the labor movement spelled trouble for the steel giant's long-term viability.[86]

Trouble also loomed for Schwab. He made a lot of money during Bethlehem Steel's heyday, but he also wasted no time spending it. Throughout his life, he was a gambler on the business front, hitting big on one venture and going bust on another. The Great Depression precipitated his personal financial demise. Shortly after the 1936 Johnstown flood, Schwab offered to sell his Riverside mansion to New York officials for $4 million, but city leaders weren't interested. Three years later, the trailblazing steel executive died in a small apartment in New York City at the age of seventy-seven. On the day of his death—September 19, 1939—he was $338,349 in debt. His body was transported from New York to Loretto and laid to rest in St. Michael's Cemetery. "I love Bethlehem," he said shortly before his death. "It is the great achievement of my life."[87]

3
THE ST. PATRICK'S DAY FLOOD

For most Johnstowners in 1936, it was a typical March. Temperatures were rising, snow was melting and the first signs of spring were emerging. Downtown, the Glosser Brothers Department Store's spring line had arrived—Steel King work shirts and Salt and Pepper Union suits were sixty-six cents each. Chambray work shirts were on sale at two for sixty-six cents. Theatergoers anxiously awaited the first showing of Charlie Chaplin's new comedy *Modern Times*, scheduled to open at Johnstown's Strand Theater on March 21. In international news, the League of Nations had rejected conditions proposed by German chancellor Adolf Hitler—conditions that he said must be met in order for his country to participate in discussions regarding the Nazis' remilitarization of the Rhineland. In a Berlin celebration, Hitler vowed that Germany's armies "can never be conquered."[88]

Despite stories that were passed down about the Great Flood of 1889 almost a half century earlier, most residents didn't think much about that catastrophic event. It was a part of history. There had been flooding incidents in the valley since then, but they were insignificant compared to what had happened when the South Fork Dam broke apart. Roughly two years after the Great Flood—on February 17, 1891—the Little Conemaugh and Stonycreek Rivers overflowed their banks. The Cambria Iron Works, as well as the Lorain Steel Company, were both forced to shut down temporarily, and streetcar service was halted until the water receded.[89]

The first Glosser Brothers Department Store opened in Johnstown in the Franklin Building in 1906. The company expanded in the 1960s with a chain of "Gee Bee" discount stores across Pennsylvania, Maryland, West Virginia and Virginia. *Johnstown Area Heritage Association.*

During the spring of 1894, the city's rivers again rose quickly following heavy rains. On the evening of May 20, 1894—in the span of an hour—both the Little Conemaugh River and the Stonycreek River rose more than six feet, washing out several miles of tracks and derailing six freight cars. Thirteen-year-old Thomas McFeeters fell into the Conemaugh River and drowned, and another teenage girl was killed when she lost her footing along the banks of the Conemaugh River and was swept away. Almost all of the businesses below Johnstown's Clinton and Main Streets suffered serious damage in the 1894 flooding, and Cambria Iron temporarily halted operations. In February 1902, melting snow in the mountains above the city caused the rivers to rise once again. Horses on Napoleon Street waded through waters that reached their underbellies, and the Haynes Street Suspension Bridge collapsed and was washed away.[90]

In the spring before the nation's financial panic in October 1907, Johnstown suffered its most significant flooding since the South Fork Dam's

In February 1891, less than two years after the Great Flood, the Cambria Iron Works was again forced to shut down temporarily because of floodwater. *Library of Congress, Prints and Photographs Division, Washington, D.C.*

rupture. On March 14, 1907, pounding rains and melting snow left the city's downtown a debris-ridden, muddy mess. The Hornerstown section bore the brunt of the damage. Homer Wressler, the teenage son of a popular pastor at the city's First United Brethren Church, was killed as he attempted to lasso some drifting logs in Hornerstown. Wressler became entangled in his rope, and water quickly enveloped him. At H.Y. Haws Livery Stable on Vine Street, a boy released several horses that swam to higher ground. When the water receded, hundreds of catfish and sucker fish lay dead in the streets.[91]

But despite these flood events, most Johnstowners believed that what had happened in 1889 was a freak occurrence. They believed that, by and large, their community was safe. They pointed to safeguards that had been put in place. The Hinckston Run Reservoir was completed in 1905, and work on the Quemahoning Dam was wrapped up in 1912. Completion of these projects gave Johnstown residents a sense of security and a reassurance that appropriate measures had been taken to protect the city from future flooding.

The winter of 1936 was one of the harshest the Johnstown region had experienced in many years, with frigid temperatures and significant snowfall. Very little of the snow that accumulated in the mountains above the city melted, as temperatures hovered well below freezing for weeks. This pattern continued through the end of February, before the mercury eventually began to rise. March settled into a predictable pattern for southern Cambria County—wet snowstorms with rain mixed in. Temperatures began rising and falling by 20 to 25 degrees Fahrenheit each day, and snowmelt saturated the ridges above Johnstown. Ice jams formed in the Stonycreek River, and city workers used dynamite to break them up, leaving slabs of ice floating in the water.[92]

From March 14 to March 17, a soupy fog blanketed the valley, rain fell steadily and temperatures spiked to 60 degrees Fahrenheit. By daybreak on Tuesday, March 17—St. Patrick's Day—about one and one-quarter inches of rain had fallen on the city. The Stonycreek River measured eight feet above base stage, and other rivers, creeks and streams were rising at a rate of around fifteen inches every half hour. By early afternoon, basements around the city began flooding, and sewers were backing up. At 2:00 p.m., Main and Locust Streets downtown were covered with water, and pools began forming on Bedford Street. Store owners hurriedly moved their inventories off of their ground floors.[93]

In Central Park, a statue of Johnstown founder Joseph Schantz was decapitated by debris in fast-moving waters that were beginning to pack a punch. Frame houses were lifted from their foundations, and cars were carried from their parking spaces. Fire department officials began deploying boats to rescue stranded residents downtown. They were taken to Johnstown's inclined plane and lifted into Westmont. The incline had been built two years after the Great Flood and was owned and operated by the Cambria Iron Works and its successors for almost forty-five years. Designed by Samuel Diescher of Pittsburgh, it included two sets of tracks and two cars that counterbalanced each another. In 1935, the incline had been purchased by the Westmont Borough for one dollar after Bethlehem Steel officials decided it was no longer needed.[94]

The incline carried more than two thousand residents to safety in March 1936. Incline supervisor Billy Grub and others worked throughout the night on Tuesday and much of the day Wednesday, lifting riders to safety. When these riders reached the top of the incline, they were taken to the Westmont Grove Pavilion and given sandwiches and coffee prepared by volunteers.[95]

Not all rescue missions were successful ones. A fifty-year-old woman and an eleven-year-old boy were killed on Vine Street when a boat they were traveling in capsized, dumping them into frigid, fast-moving waters that swept them away. Dan Gallagher, a streetcar operator, scaled a gas pump

The city's inclined plane is 896 feet long and ascends 502 feet to the top of Yoder Hill in Westmont. *Johnstown Area Heritage Association.*

at a filling station at the corner of Main and Jackson Streets in an attempt to escape the rising waters. He lost his balance and drowned. Andrew Tkac survived the flood but not before having a harrowing experience. The eighteen-year-old was standing in the living room of his family's home on Chestnut Street in Cambria City, a room that was already under several feet of water, when a strong current rushed in and swept him out the front door. Tkac grabbed hold of a telegraph pole that floated within his grasp, but the swift-moving waters carried him and the pole into the Conemaugh River. The pole then slammed into the city's Ten-Acre Bridge. Tkac managed to grab a girder and pull himself out of the water, which was level with the floor of the bridge. Soaking wet and freezing cold, he climbed the girder all the way to the top, staying there until the following morning, when the water receded. Debris crashed into the bridge throughout the night, but it did not break apart. "I felt very lonely," said Tkac.[96]

Mine foreman T.P. Bradley's quick response saved the lives of members of his crew. After water in the Benscreek River began pouring into the Hughes No. 2 Mine in the village of Cassandra, Bradley rushed into it and alerted eighteen of his miners, who quickly evacuated. There were peculiar

Vine Street looked more like a river by midafternoon on St. Patrick's Day in 1936. *Johnstown Area Heritage Association.*

Property losses for residents of Baumer Street continued to rise as the floodwater rose.
Johnstown Area Heritage Association.

sights amid the destruction. Baby grand and upright pianos floated in the floodwaters near Franklin Street early Wednesday morning. The pianos had been swept out of the Capitol Building, where they had been on display hours earlier. On Washington Street, mannequins floated in the floodwaters, creating an eerie scene.[97]

By late Tuesday afternoon, most of the phone lines across the city had stopped working. Between noon and 10:00 p.m., three more inches of rain fell, creating more problems. People became trapped at different locations around the city, including the Cambria Theater, the Glosser Brothers Department Store, the Penn Traffic Store and the Snook Hardware Store. The Maple Avenue, Franklin Street and Poplar Street Bridges washed out. Repairs to these and other damaged bridges cost more than $1 million.[98]

By nightfall, one-third of the city was under water. In the Cambria City section, the waters reached heights higher than those recorded during the Great Flood. Around 9:30 p.m., a rumor began circulating in parts of the city that sent many into a panic. A police officer driving through the Seventh Ward and Dale Borough warned residents that a dam had broken. The officer told residents to head for higher ground immediately. The claim was untrue, and no dam had failed. "Men, women and children, babies in arms, carried what they could, scrambling to the hills of Dale," noted a story in the *Tribune*.[99]

Pianos that had been on display in the Capitol Building floated along Franklin Street before coming to rest in mud and standing water. *Johnstown Area Heritage Association.*

The metal truss Franklin Street Bridge floated along the Stonycreek River during and after the flood, much like the boats that carried rescue workers to various locations throughout the city. *Johnstown Area Heritage Association.*

The Little Conemaugh River crested around 7:30 p.m. on Tuesday. The Stonycreek escaped its banks at about 11:30 p.m. An hour later, the Conemaugh River spilled out of its home.[100]

JOHNSTOWNERS CLEAN UP AND REBUILD AGAIN

Sunlight gleamed brilliantly early Wednesday morning in a cloudless sky above a city devastated by floodwaters for a second time. Early that afternoon, however, the skies clouded over once again, and the rain resumed. The rumor of a dam breaking also resurfaced. A man from the community of Jerome, just south of Johnstown, transported his pregnant sister to Memorial Hospital that morning. As he was heading home, he heard a disc jockey on his car radio discussing the rumor of a dam break. He turned his car around, drove back to the hospital and demanded his sister be discharged. The two then drove to Westmont. When the rumor proved unfounded, the man drove his sister back to Memorial Hospital. Ten minutes after her second hospital admission of the day, the man's sister delivered her baby.[101]

"Pandemonium reigned again," noted a *Tribune* reporter. "Thousands made a rush for Frankstown Road, the Prospect section and the Westmont hillside. In Ferndale, police sounded the sirens, and people fled to higher ground. Thousands were soaked to the skin as they stood on the hillsides, watching for the rush of water that was supposed to be on its way down the Stonycreek Valley. But the flood did not come, and for the second time, the people returned to their homes."[102]

On Friday morning, Johnstown mayor Daniel Shields assured residents that all of the dams across the region were stable. "I have had all of them inspected by engineers," he said. The mayor said that water in the dams had not approached the tops of their breasts and that no cracks had developed. "All of the dams are in excellent condition," echoed Charles Kunkle, vice-president and general manager of the city's water company.[103]

Ninety-year-old J.C. Beaner and his son, Charles, fled their home on Pine Street and headed for higher ground above the city when the rumor of a dam break swept through town for a second time. Forty-seven years earlier, when the South Fork Dam gave way, Beaner served on Johnstown's seven-member police force. During the Great Flood, he swam several blocks with his then-toddler son, Charles. Eventually, the father and son found a safe spot above the water. Beaner worked without sleep for two straight days and nights in 1889, rescuing trapped residents.

The 1936 flood left the Beaners' house in ruins. J.C. Beaner wanted to help his city rebuild and clean up for a second time. His son told him he would have to sit this flood cleanup out. "I wanted to help," said the elder Beaner, "but they said I was too old." The younger Beaner sent his father to Akron, Ohio, where he stayed with his daughter. J.C. Beaner returned to Johnstown as soon as serious risk of disease had passed. "This is my home," he said.[104]

Twenty-four people died in the 1936 Johnstown flood. Seventy-seven buildings were destroyed, nearly three thousand more were severely compromised and the total property damage was estimated to cost around $41 million. The economic toll on downtown businesses was severe. Nat Shendow's Men's Shop reported losses of $11,000, and at Lord's Dress Shop, damages totaled $20,000. Officials at the Penn Traffic Company estimated losses of $450,000. "Everything that was under water, or in close proximity of the water, had to be disposed of," said Mike Wolfe, manager of the Penn Traffic location downtown. "We just took it out to Washington Street, which was blocked off, and had trucks haul the debris away." The Penn Traffic Store enjoyed a long history in the city. It first opened in 1854 as Stiles, Allen, and Company and served as the company store for the Cambria Iron Works. It changed its name in 1855 to Wood, Morrell, and Company, before becoming Penn Traffic in 1891. For many years, it was the largest store in the city.[105]

At the Johnstown Poster Advertising Company, General Manager Tom Nokes estimated losses of $50,000 to $60,000. The John Thomas and Sons Department Store suffered losses in excess of $100,000, and the Kinney Shoe Shop sustained $75,000 in losses. Rothstein's Jewelry Store saw $50,000 worth of diamonds and watches destroyed. Many of these businesses' losses were covered by flood insurance protection, but not all. "Johnstown merchants were in readiness for the full swing of spring trade, and in nearly every instance, they had already received their full stocks of merchandise," said one downtown store owner. "Showrooms were completely stocked with new merchandise, and display windows were made ready, at a large outlay of money, for the formal spring openings."[106]

Mayor Shields, who had held the position for just two months before the St. Patrick's Day flood, was widely praised for his handling of the crisis. He was elected the city's mayor in the fall of 1935, thanks, in large part, to financial support for his campaign from officials at Bethlehem Steel. A lifelong resident of the region and a former foreman at Bethlehem, Shields had close ties with L.R. Custer, Bethlehem's general manager. A year after

Losses from the 1936 flood at the Penn Traffic Store surpassed $450,000. *Johnstown Area Heritage Association.*

the flood, Shields played an instrumental role in helping Bethlehem officials secure a victory over striking laborers.[107]

After electricity across most of the city had been knocked out—and after witnessing prominent Johnstown photographer Frank Buchanan drown in floodwaters in front of his studio in the Otto Building on Franklin Street—Shields, on Custer's invitation, established an emergency headquarters at Bethlehem's offices on Locust Street. Bethlehem's power system had remained intact during the flooding, and its Locust Street location was one of the few buildings downtown that had uninterrupted electricity on Tuesday night. Bethlehem's telegraph and telephone lines were down that evening, but both services were restored by early Wednesday.[108]

Pennsylvania governor George Earle flew over Johnstown by helicopter on Wednesday morning and inspected the devastation. He dispatched 1,838 members of the Pennsylvania National Guard to the city. These troops were brought into the valley on trucks, as rail travel was impossible—miles of tracks had been washed away by the flooding. One hundred Pennsylvania State Police officers and eighty highway patrolmen from across the state also arrived on Wednesday.[109]

On Main Street and throughout the downtown section of the city, the floodwater damaged or ruined much of the inventory of various businesses. *Johnstown Area Heritage Association.*

With the exception of residents getting to and from their homes and truck drivers transporting food, drinking water and medical supplies, anyone else who wanted to enter the city needed a pass. "We are here to assist Mayor Daniel J. Shields in enforcing the law and protecting property," said Colonel C.B. Smathers, who oversaw the national guard troops. "We will take stern measures to carry out our efforts if it becomes necessary to do so." Any person who could not prove residency or show legitimate reason for entering the city was turned away. "The major problem, aside from that of restoring the city from the ravages of the water," noted an editor at the *Tribune*, "has been that of keeping the affected area free from the morbidly curious sightseers who flocked to Johnstown to witness the effects of the flood. An intolerable condition existed until the guardsmen and state police got their forces properly organized and restricted zones were established."[110]

On Wednesday morning, Shields ordered all businesses that sold alcohol—including state-operated liquor stores—closed until further notice. When the magnitude of the crisis became clear, the Pennsylvania Liquor Control Board gave local authorities in flood-stricken areas the power to determine whether or not beer and liquor could be sold. Shields also ordered all of the city's banks to close for an indefinite period, and he instituted a 9:00 p.m. curfew. "The city of Johnstown is face to face with a catastrophe, the size of which cannot be exaggerated," said the mayor. "I believe the city will arise from the flood greater than ever. However, in order to do so, federal and city officials must have the complete cooperation of every man, woman and child. Anyone making a false move will be dealt with severely. Crooks from outside who believe they may reap a harvest by coming to Johnstown to loot had better stay away."[111]

A group of men from Somerset, a community thirty minutes south of the city, trucked in loaves of bread to Johnstown, where they sold them for around forty cents a loaf, well above the average price at the time. Exploitation of the tragedy was much more the exception, however, than the rule. As was the case in 1889, community members rallied to help their neighbors.[112]

Somerset County potato farmers contributed parts of their harvest from the previous fall. Altoona American Legion Post 228 and the Altoona Boosters Association delivered 5 tons of clothing and bedding, 1 ton of

Daniel Shields, a Republican, served as the mayor of Johnstown from 1936 to 1939. *Library of Congress, Prints and Photographs Division, Washington, D.C., photograph by Harris & Ewing.*

Robert Bondy, the director of disaster relief for the American Red Cross, flew over Johnstown following the flood and dispatched Red Cross workers to the city shortly thereafter. *Library of Congress, Prints and Photographs Division, Washington, D.C., American National Red Cross Collection.*

medical supplies, 5,000 loaves of bread, 250 gallons of milk, sandwiches and plenty of coffee. The Blair County Milk Dealers Association delivered truckloads of butter, eggs and milk. The Barnesboro community in the northern part of Cambria County sent 10 cases of milk, 200 pounds of coffee, 40 bushels of potatoes, 200 pounds of beans and 400 gallons of gas. Hundreds of truck owners volunteered their time and vehicles to transport supplies and haul away debris. Fire companies from throughout the region arrived in Johnstown on Thursday afternoon and began pumping water from the basements of downtown businesses and homes.[113]

Conemaugh Valley Memorial Hospital, located in the city's Eighth Ward, opened its doors to displaced residents. Hospital staff housed and fed members of the police force and Red Cross. "Our employees are working day and night," said Herbert Fritz, the hospital's superintendent. "This hospital was built on the surplus funds of the flood of 1889, and we now have an opportunity to show our appreciation." Reverend John Codovi, vicar general of the Roman-Catholic Diocese of the Altoona-Johnstown region, announced that the church was suspending all Lenten season fasts for those in flood-stricken areas, as well as other communities that relied on Johnstown for their food supply.[114]

Robert Bondy, director of disaster relief for the Red Cross, flew into Johnstown on Wednesday and assigned twenty-two members of the organization's national staff to the city. They used the first floor of the Grand Army Memorial Hall as their headquarters and organized eighteen refugee centers around Johnstown to feed, clothe and shelter survivors. At two of these centers in the Daisytown section of the city, six hundred people sought shelter. The Red Cross also set up seven first-aid stations and twenty-one coal distribution sites. At the coal sites, area mining companies dumped truckloads of coal and invited families to take what they needed to heat their homes. "Hundreds of persons are doing admirable work in this crisis," said Bondy. "The response of volunteers has been outstanding and is one

WPA workers who arrived in the city helped with the cleanup in many ways, among them, the removal of mud, which had accumulated seemingly everywhere, including in front of the city's train station. *Johnstown Area Heritage Association.*

of the most gratifying features. Although they have suffered heavy losses themselves, many persons have taken refugees into their homes. With this spirit prevailing, Johnstown will come back and come back strong."[115]

The spread of disease was one of the greatest fears following the 1936 deluge, just as it had been in 1889. Many animals died in the floodwaters, and volunteers and city employees worked quickly to clean up their remains. "The medical forces of the city and county, augmented by the state public health service, have the situation well in control," said Dr. W. Frederick Mayer, president of the Cambria County Medical Society. "There is no typhoid fever or other forms of contagious or infectious diseases, although the medical forces are entirely organized to cope with any threats from this source."[116]

Members of the Works Progress Administration—one of President Franklin D. Roosevelt's New Deal initiatives—also provided assistance. Around 7,000 WPA workers and 350 trucks were sent to the city once the water began receding. The WPA assisted in cleaning up the overwhelming mess, dumping tons of mud into the rivers. These men also burned animal remains and spoiled food outside of the city and helped rebuild washed-out roads and bridges.

POACHING SEASON

Following the flood, several of Johnstown's political leaders worried that businesses would flee the city due to the risk of future flooding. Their fears were well founded: a number of leaders in neighboring cities and states looked to poach businesses from the valley. North Shaver, vice-president of Johnstown's Penn Machine Company, received the following correspondence from William Eaton, executive vice-president of the Canton, Ohio Development Corporation:

Flood conditions raise factory production costs—no argument there. How much would you have saved if you could have carried full insurance against flood damage? Are your competitors in more favorable position today because they were not flooded? Will you have to charge increased costs because of the expense of replacing or reconditioning buildings or equipment, because of interrupted production, interrupted transportation, light or water service, disrupted personnel, and delayed shipment?

The only real flood insurance is a flood-free location. Canton, Ohio, has no flood damage whatever—now or ever. We would like to present for your consideration more detailed information on the advantages, fitted to your special needs, which you would secure by locating all or part of your operations in the Flood-Free Canton Industrial District. May we hear from you?

Yours sincerely,
Wm. W. Eaton

Eaton received return correspondence from Shaver several weeks later:

We are in receipt of your letter of March 26 outlining the many advantages of factory location in Canton. In spite of the fact this company suffered heavy property damage in the recent disaster and have no strings to keep us here, we have no intention of leaving—we like it too well. If the water gets too troublesome in the future, we can move our plant to one of the many beautiful hills that have only recently been found so convenient and comfortable.

Yours very truly,
North C. Shaver

An editor at the *Tribune* caught wind of the solicitation letter that Eaton sent to Shaver and slammed his effort to exploit the tragedy: "In their zeal to increase industrial payrolls and thus add to their material prosperity, some community organizations resemble ghouls robbing a graveyard. They lose all sense of fairness and common decency in their efforts to achieve a record of industrial expansion."[117]

Bethlehem Steel's Charles Schwab assured Mayor Shields that his company was staying in Johnstown. Schwab told Shields that Bethlehem was actually planning to invest in improvements to its mills in the region in the wake of the flood. A Hiram Swank and Sons advertisement noted:

> *Having been a Johnstown institution since way back in 1856, we've known the city in good times and bad, in fair weather, flood, and fire. Just forty-seven years ago, our plant, then located on the Southside, was wiped out by the flood of 1889. Yet never in all these years have we felt as proud of the courage and fortitude of our people—our friends, neighbors, and business associates—as we do today, following our recent flood. Such remarkable recovery indicates a remarkable people.*[118]

FEDERAL GOVERNMENT GETS INVOLVED

While Schwab's promise that Bethlehem Steel would rebuild and even expand its operations in the valley was enthusiastically received by city officials, Johnstowners needed more help. City residents mailed more than fifteen thousand letters to the White House, pleading for federal assistance. An editor at the *Tribune* summed up the challenges the city faced:

> *The forces of nature are entirely beyond our control. Nothing we may do will prevent abnormal rainfalls nor limit the amount of snow that descends upon our mountains. We shall continue to witness both at such intervals as it shall please the Almighty to send them. However, it is within the power of man to establish safeguards that will minimize the danger and reduce the losses.…The cost* [of flood protection] *should be borne by the federal government and the work done under the direction of the public works administration. Diversion of federal funds from trifling projects conceived solely for the purpose of creating work for the unemployed will make this possible without any sacrifice of employment for the needy.*[119]

Left to right: Pennsylvania governor George Earle (*pointing*), President Franklin D. Roosevelt and Johnstown mayor Daniel Shields during Roosevelt's visit to the city on August 13, 1936. *Johnstown Area Heritage Association.*

On August 13, 1936, five months following the flood, President Roosevelt visited Johnstown. "I came here to see conditions with my own eyes," said Roosevelt. "I believe I can render better service after getting a firsthand view than if I just stayed in Washington." He arrived in the city via railcar in the late afternoon. Andrew Shields, Mayor Shields's son, drove Roosevelt, Pennsylvania governor George Earle and two officers from the U.S. Army Corps of Engineers (W.E.R. Covell and Edward Markham) around the city in Schwab's Packard limousine. The group toured Johnstown for two hours and visited several potential dam construction sites before Roosevelt delivered a speech at Roxbury Park. Willard Edwards of the *Chicago Tribune* documented Roosevelt's visit:

> *The speed of the presidential special, which left Washington at 10 o'clock this morning, was purposely slowed during the afternoon so that the party arrived in Johnstown at 5:00 p.m., when the streets were full of workers released for the day. More than 30,000 of them lined the*

streets and gave the president tremendous applause as he rode in an open car through the streets.

The rally in a large park tonight was about as nonpolitical as a Fourth of July picnic given under the auspices of a ward committeeman. Elaborate loud-speaking apparatus had been set up. Both Gov. Earle and Sen. [Joseph] Guffey poured into the microphone eulogies of President Roosevelt as the only executive who ever had paid any attention to the sufferers in flood areas, and President Roosevelt then followed with his promise to see that Johnstown gets its share of the $315 million authorized by the last Congress when the time comes to spend the money on flood control projects. The president pictured Johnstown citizens as living in constant peril and roused cheers with a tribute to his listeners for their courage.

"You good people have shown the highest qualities of good American citizens," he declared, "and as long as I have anything to do with the government, I am going to see that you are protected from floods."[120]

The city's channel system and river wall construction was inspected and approved on November 27, 1943. The project required the excavation of 2.75 million cubic yards of soil. *Johnstown Area Heritage Association.*

On August 28, 1937, Roosevelt signed the Omnibus Flood Control Act. "Flood protection shall be provided for said city [Johnstown] by channel enlargement or other works," noted the president. The U.S. Army Corps of Engineers began channelizing sections of the Little Conemaugh and Stonycreek Rivers in August 1938. The purpose of this $8.7 million "Local Flood Protection Program" was to increase the capacity of the rivers. The project was inspected and approved on November 27, 1943. Included in the project was the construction of 6,500 feet of river walls and 3,000 feet of dykes. Corps members excavated 2.75 million cubic yards of soil during construction. James Bogardus, Pennsylvania's secretary of forest and waters, did not believe that these flood protection measures were sufficient. He recommended that three reservoirs be constructed within tributaries of the Stonycreek River in addition to the channel system. These measures, however, were not pursued.[121]

The channel system was designed to prevent a flood equal to that of the 1936 deluge, but it was not designed to contain a flood that met "standard project flood" (SPF) levels. SPF is defined as a "flood that may be expected from the most severe combination of meteorological and hydrological conditions considered reasonably characteristic of the geographical area in which the drainage basin is located." Following completion of the channel system, President Roosevelt wrote to Walter Krebs, chairman of the Flood-Free Johnstown Committee: "Johnstown, from now onwards, will be free from the menace of floods. Happily, for the future of Johnstown, its citizens can now devote all their energies to their ordinary pursuits, without worry over the impending hazard of uncontrolled waters."[122]

First Lady Eleanor Roosevelt made a trip to Johnstown in November 1940:

I visited the Red Cross Roll Call Headquarters and saw some of the garments which they are turning out and shipping to Great Britain and Finland. Then I went with the mayor to see the flood control work which the federal government is helping them to put through.

Two terrible floods have visited Johnstown. The one in 1889 cost the city a great many lives. Seven hundred and seventy three people were never identified, and up in the cemetery on the hill, the unmarked stones are placed in rows as they are in Arlington Cemetery in Washington. One monument is erected to all those unidentified dead. The recent flood cost the city some $40 million, but fortunately, only a few lives were lost in the whole county.

President Roosevelt and other elected officials assured Johnstowners that their city was "flood-free" following the completion of the channel system in 1943. It was—until 1977. *Johnstown Area Heritage Association.*

The work of dredging the two rivers, which will, in the future, safeguard the population from the ravages of the past, is in full swing, and I was much interested to see it. The president inspected this work and told me of it some time ago, so I know he will be glad to hear of the progress made.

It is easy to see that a city, which has suffered as this one has, must take advantage of whatever help the state and federal government can give. For the future, these expenditures to prevent recurring disasters will be a saving not only to the community but to the state and the nation.[123]

Johnstown was not the only community to suffer catastrophic flooding in March 1936. Thirteen states across the Mid-Atlantic and New England regions experienced serious flood events that month. Pittsburgh suffered the worst flooding in its history, with water levels reaching forty-six feet in some locations. Sixty-nine Pittsburgh residents were killed and property loss in the city was estimated to cost $250 million. Steel mills that operated near the confluence of Pittsburgh's three rivers—the Monongahela, Allegheny and Ohio Rivers—suffered significant damage. In Maryland and West Virginia, the Potomac and James Rivers spilled over their banks, causing serious flooding. In New England, the Connecticut River reached flood stage, and twenty-eight people died in floodwaters in Connecticut. Major flooding also occurred in New Hampshire, where the Merrimack River crested above eighteen feet.

In April 1937—a little more than a year after the 1936 St. Patrick's Day flood—Johnstown again experienced serious flooding. The city's rivers spilled over their banks yet again following two days of nonstop rain, filling parts of the downtown with four to five feet of water. A temporary bridge on Franklin Street—one that had been built by WPA workers—washed away, and Bethlehem Steel was forced to shut down operations at several of its mills. Just before the rain stopped, weather forecaster Robert Tross noted that "if another half inch of rain falls in the next five hours, it will be bad."[124]

In one of the most unusual stories from Johnstown's 1936 St. Patrick's Day flood, a woman reclaimed her wedding ring, one that had been missing for twenty-eight years. Mrs. M.D. Helsel and her husband had settled into their home in Johnstown after marrying in 1906. Shortly thereafter, the couple moved to nearby Holsopple. Their original home in the city changed hands several times and was owned by Charles Zhrozek at the time of the 1936 flood. As the floodwater receded, Zhrozek noticed something shiny in the garden in his backyard, where much of the topsoil had been washed away. He picked up the object and discovered that it was a wedding ring engraved with the message "M.D.H. to T.E.A." Following some detective work, Zhrozek located Helsel and returned the ring to her.[125]

4
WAR, PEACE AND NEGOTIATIONS

Shortly before 8:00 a.m. on December 7, 1941, Japanese fighter pilots bombed the U.S. naval base in Pearl Harbor, near Honolulu, Hawaii. More than 2,400 Americans were killed and over 1,000 were wounded in the surprise assault. Johnstowners learned of the attack in the early afternoon of December 7. Johnstown mayor John Conway—who had succeeded Daniel Shields—told residents, "We're going to put Johnstown defenses in order to immediately protect our people. The mills are our first task. If anything happens there that doesn't look right, we're prepared to move at once."[126]

Pennsylvania governor Arthur James told Conway and several other mayors of cities around the state to be on guard for potential attack by the Japanese. In Johnstown, the governor ordered the following sites to be placed under round-the-clock surveillance: Bethlehem Steel, the Johnstown Traction Company, the Johnstown Water Company, the Lorain Division of U.S. Steel, the National Radiator Corporation, the Pennsylvania Electric Company, the WJAC Radio Station and the Western Union Telegraph Company.[127]

Johnstown's importance in any U.S. military buildup was obvious, and Governor James believed that the city's legacy of steel production made it an attractive target to Axis leaders. Former U.S. civil defense director Fiorello LaGuardia visited Johnstown shortly after the Pearl Harbor attack. "The Germans are pretty aware of Johnstown and its steel mills," said LaGuardia. "It probably would be one of the central targets in an attack."[128]

Bethlehem Steel's plants were among several different locations in Johnstown on high alert following the Japanese air raid of Pearl Harbor. *Library of Congress, Prints and Photographs Division, Washington, D.C.*

The munitions plant in the Coopersdale section of Johnstown forged hundreds of thousands of shells for the Allied powers during World War II. *Johnstown Area Heritage Association.*

Johnstown's Bethlehem Steel Plant, as well as the company's other holdings, proved critical to the war effort. The company's shipbuilding division constructed 1,127 vessels during the war. Bethlehem president Eugene Grace promised President Roosevelt that his company would produce a ship a day during the war—it exceeded this commitment by fifteen ships. Bethlehem plants produced parts for bombs (guide fins), guns (artillery forgings, breech-ring forgings, breech blocks, torpedo heads), ships (diesel engine heads, turbine rotors) and tanks (brake drums, flywheels). Bethlehem officials also oversaw operations at a munitions plant in the Coopersdale section of Johnstown that was built by the federal government. The $4 million plant began producing Howitzer shells in February 1945.[129]

"War needs ships," said Grace. "The United States' production of ships has been a prime factor in turning the tide to victory. There's no other navy afloat that, ship for ship, man for man, gun for gun, can equal the United States' fleet. And we are proud to have had a major part in the building of this magnificent fighting force."[130]

With many of the company's employees enlisting for military service, Bethlehem launched a major recruitment initiative. "Our employment people were beating the bushes for miles around, trying to find people to fill the gap," said Eugene Simmers, a superintendent of heat treatment and forging at Bethlehem during the 1940s. "We trained everyone we could get ahold of. Eventually, we went to hiring women, training them to fill jobs that had historically been done by men."[131]

On the evening of August 14, 1945, President Harry Truman announced to the nation that the Japanese had surrendered. World War II was over, and around the country, people took to the streets to celebrate. A "victory bell" tolled in Westmont, and the headline in the *Tribune* the following morning read, "World Enters New Era of Peace."[132]

More than 405,000 Americans were killed in the fighting during World War II, and Johnstown, like many other working-class communities across the United States, paid a heavy price in the effort to defeat the Axis powers. The Cambria County War Memorial in the downtown section of the city lists the names of 858 area residents who were killed in the fighting.

THE RISE OF ORGANIZED LABOR

By the end of 1945, roughly forty-three thousand servicemen from Cambria and neighboring Bedford, Blair and Somerset Counties had been discharged from military service. At Bethlehem Steel in December 1945, Grace stepped down as president, and Art Homer was named chief executive officer. Grace continued to serve as chairman of the board at the company until 1957. One of Homer's first orders of business as CEO was to trim the salaries of the company's executives, which had ballooned under Grace's watch. In 1929, Grace ranked as the highest paid business executive in the United States, receiving $1.6 million in compensation and bonuses. Homer also announced a modernization project for Bethlehem's Johnstown operations, which employed more than eighteen thousand workers in 1945 and produced 2.25 million tons of steel ingots annually. Open-hearth furnaces in the city were modernized, and improved cranes and ingot buggies were added.[133]

The quality of life for many steelworkers and coal miners in Johnstown improved following World War II. This was thanks, in part, to union leaders, who concentrated their organizing efforts on the Mid-Atlantic region after the war. According to labor historian Jack Metzger:

> *In Johnstown, the United Steelworkers organized everyone in sight, from small machine shops to dairy workers. The local press, politicians and business leaders gradually learned a cautious respect for blue-collar workers, whether they were in a union or not. The USWA not only established the standard for wages and benefits in the area, it set the social tone for class status as well. Steelworkers learned that merchants, doctors and schoolteachers respected a man with a couple of bucks in his pocket. Steelworkers' wages and benefits and steelworkers' power were at the root of everyone's improved condition of life and prospects for the future.[134]*

Union organizers' success in improving the lives of laborers and their families was not limited to securing higher wages. "Every time we got a wage increase, everything else went up. So, what did we gain by that?" said Frank Guidon, who joined Bethlehem Steel in 1936. Guidon worked a number of different jobs with the company, including as a table operator, guide setter, greaser, screw-down operator and roller. "But all of the other benefits that we got, like pensions and Blue Cross-Blue Shield—the company paid for all that stuff—that worked out pretty good. Vacations was another thing. By the time I ended at the company, I was getting thirteen weeks [per year]."[135]

The labor movement also led to a decline in some of the longstanding and exploitative practices by members of management. "Before the unions came in, the usual procedure was you worked all three shifts, unless you had a farm and you brought in potatoes or something to the foreman, and then you got a pay-shift job," said Herbert Sechler, who worked for Bethlehem as a machinist from 1928 to 1974. "We used to hang our coats on a nail on the wall as soon as you came in the door. Some of the foremen would go along the coat and feel to see what was in the pockets. Those are the guys that got work. And complaining about it didn't do any good."[136]

Union organizers often intercepted mill workers and coal miners at the beginning and end of their work days, urging them to join the ranks of organized labor. *Johnstown Area Heritage Association.*

Members of management weren't the only ones who employed heavy-handed tactics with laborers. Union leaders bullied Bethlehem's workers in order to get them to join their ranks. John Whitney started at Bethlehem in 1941 and remained on the company's payroll for thirty years:

> *You would run into a* [union man] *on the way in, and you were asked, in a nice way, if you wanted to join. And if the answer was in the negative, some of them boys needed first aid. At first, I didn't belong to the union, and twice, I was approached. And they looked like some of these wrestlers that you see on TV. The second time, I had one on each side of me and they said, "How about it, buddy, you gonna join up?" And I'll never forget, I was in a sweat. I didn't want to get in, but I also didn't want to take a lacing. They were big boys, and they could have given me a good polishing. I said, "The hell with you, the hell with the steel company. I'm for me. Next week, I'm going in the service. And not you and not the company are going to worry about me." And I got away with the bluff.*

Whitney eventually joined the union in 1951. "I decided that I couldn't buck the tide forever, and the union was doing stuff for us. I didn't want to be a freeloader."[137]

STEEL AND COAL'S POSTWAR DECLINE

Like steel production, coal mining also declined in Johnstown and other manufacturing cities in the years following World War II. Mining companies in Cambria and Somerset Counties employed 20,275 workers in 1950 and produced 12.5 million tons of coal that year. Ten years later, the number of people employed by mining companies in the two counties plummeted to 6,560, and coal production in these counties had dropped to 8.9 million tons. In 1950, 145 coal companies operated in Cambria County. By 1960, 6 remained in operation. A similar dramatic downsizing of the mining industry was seen across the nation.[138]

There are a number of factors that contributed to this decline. More and more Americans were heating their homes with natural gas instead of coal. In the railroad industry, locomotives that had once been powered by coal ran on diesel. Within the steel industry, technological advances allowed companies to use much less coal in the production process. The mining industry was also becoming increasingly mechanized, allowing companies to extract large amounts of coal from underground seams with far fewer miners. Some companies shifted to more strip mining and less underground extraction. And while markets for coal were decreasing, competition from other coal-producing states was increasing. The convergence of these factors resulted in the elimination of thousands of mining jobs around the Johnstown region.[139]

At Bethlehem Steel, increased international competition following the war exacted a serious toll on the company's profit margin. In 1950, American plants produced 46 percent of the world's steel. By 1970, only 20 percent of the steel being used around the world was made in the United States. In 1955, the United States imported approximately 1 million tons of iron and steel; by 1968, roughly 18 million tons entered the country annually.[140]

Another reason for Bethlehem's downturn after World War II was its failure to innovate. The company's mills were aging, and few large-scale capital improvements had been pursued in the post-Schwab era. It is also important to note that steel was not being used nearly as much as it had been in the past. Many products following World War II were being produced with aluminum, ceramics and plastics. Increased concerns about the environment also played a role in Bethlehem's—and the overall U.S. steel industry's—decline. The Environmental Protection Agency began placing restrictions on pollution emissions, requiring steel companies to comply with more rigid regulations. Companies were given deadlines to meet these new standards. The way steel was made was also changing. Open-hearth furnaces gave way to basic oxygen furnaces (BOFs). Invented in Austria in 1952 and introduced to the United States two years later, these furnaces blew oxygen through molten pig iron, lowering the carbon content in the steel they produced. BOFs produced steel faster and required less manpower than open-hearth furnaces. Plus, they were less harmful to the environment.[141]

The most significant factor in the decline of Bethlehem Steel over the second half of the twentieth century, however, was the rise of organized labor. Charles Schwab, Eugene Grace and other Bethlehem executives had worked aggressively—and successfully—for many years to keep workers from unionizing, but labor organizers had gained the upper hand following World War II. Bethlehem began paying its workers higher and unsustainable wages while also offering them more expensive benefit programs, improvements negotiated by the unions. These wage increases and benefit packages exacted a serious toll on the company's bottom line—a toll that, in the end, would be too much for the company to overcome.

"You watch the program *The Jeffersons*—everybody wants a piece of the pie?" said George Dancho, who first joined Bethlehem in 1936. Dancho left the company for military service during World War II and returned to Bethlehem after the war. He worked as a window cleaner, night watchman, roller and guide setter. "We got a piece of the pie. Steel isn't right all the time, and the union isn't right all the time, but we sure got a piece of the pie."[142]

STRIKE OF 1959

The year 1959 will live in infamy for Bethlehem Steel and the American steel industry. It was the year that industry executives joined forces in an effort to reclaim some of the ground lost to organized labor. Steel leaders decided to negotiate as a coalition, taking a page from union organizers in hopes that a consolidated front could work similarly for them. It didn't.

The steel companies' lead negotiator was R. Conrad Cooper, the son of a Kentucky coal miner and a former football star at the University of Minnesota. "It is time to call a halt to twenty years of inflationary excesses that have placed the steel industry under a serious competitive handicap," said Cooper. Along with U.S. Steel chairman Roger Blough, Cooper met with President Dwight Eisenhower. Cooper asked the president not to intervene on organized labor's behalf but to instead let the two sides handle the situation among themselves. He told Eisenhower that the steel industry was collapsing under the weight of salary and benefit gains negotiated by the unions and that his side was prepared for a prolonged fight to reset the balance of power in the industry. Eisenhower promised Cooper and Blough that he would stay out of the fray.[143]

Cooper then met with United Steelworkers Association president David McDonald and demanded that the wages of union members be frozen for a year. Cooper also told McDonald that revisions had to be made to something known as "Clause 2B." Adopted in a collective bargaining agreement between management and labor in 1956, Clause 2B limited management's ability to change the number of workers assigned to a task or to introduce new work rules or machinery that would result in reduced hours for workers or fewer workers on the job. While some labor historians believe that Clause 2B was not a significant factor in the decline of the American steel industry, its impact on the ultimate demise of Bethlehem Steel, specifically, was consequential.[144]

The term "featherbedding" was synonymous with Clause 2B. This practice involved hiring more workers than were needed to perform a job, as well as assigning meaningless and time-consuming tasks to employ additional workers. Featherbedding was common practice at Bethlehem. According to historian John Strohmeyer:

> *Self-interest kept* [Bethlehem] *supervisors from stepping forth and providing the facts that could have knocked out Clause 2B. Plant bosses chose to protect their workforce for the day that production would pick back up again. And who could blame them? The prevailing attitude had always been, "Let's stay ready for the big years." No one on top was telling the plant bosses to think differently.*[145]

McDonald told Cooper that the union was not opposed to technological improvements but that Clause 2B needed to be maintained. For Cooper and the steel leaders he represented, getting rid of Clause 2B was a line in the sand. The two sides were at a stalemate. On July 15, 1959, 500,000 steelworkers across the country went on strike. On September 29, Eisenhower met with McDonald and the steelworkers' general counsel Arthur Goldberg. The president told the two men to reach some sort of compromise or he would invoke the Taft-Hartley Act and force workers back into the mills.[146]

Despite Eisenhower's demand, Bethlehem officials and most of the nation's other steel leaders refused to budge on their stance that Clause 2B had to go. On October 20, citing the Taft-Hartley Act, the Department of Justice petitioned the federal district court in western Pennsylvania to order the striking steelworkers in Johnstown and the surrounding region to return to work. Goldberg argued that the Taft-Hartley Act was unconstitutional, but the district court disagreed. The union appealed the ruling to the state's Third Circuit Court of Appeals and lost on October 27. The union took the case to the U.S. Supreme Court and lost again on November 7. The constitutionality of the Taft-Hartley Act was upheld.[147]

While management won this battle, it would be the unions that ultimately won the war. In December 1959, Vice President Richard Nixon met privately with the nation's steel leaders and told these men—among them, Bethlehem CEO Art Homer—that the U.S. Congress would soon open hearings to investigate the strike. Nixon also told the assembled steel bosses that the prolabor Democratic majority in Congress would cast them in a very poor light if the lingering disagreement between management and union officials produced a recession. This was an increasing likelihood given that 85 percent of the American steel industry had sat idle for almost four

BETHLEHEM REVIEW
SEPTEMBER, 1959

The Strike
for More Inflation

Left: The 116-day strike in 1959 was the longest steel strike in U.S. history and cost more than $6 billion in wages and production losses. The strike shut down 87 percent of the nation's steelmaking capacity. *Courtesy of the National Museum of Industrial History. All rights reserved.*

Below: Vice President Richard Nixon (*center*) met with steel officials in late 1959 and urged them to reach a settlement with labor leaders in order to get the nation's mills running again. *Library of Congress, Prints and Photographs Division, Washington, D.C.*

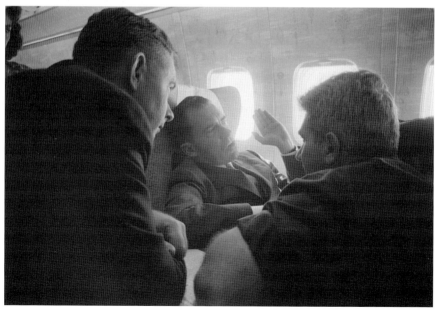

months during the standoff. Nixon urged the steel leaders to reach some sort of agreement with union officials.

On January 20, 1960, the two sides agreed to a new contract, one that preserved Clause 2B. The contract also provided workers with a cost-of-living wage adjustment and improved pension and health benefits. Before this agreement, workers and management had contributed equally to these packages. Now, Bethlehem and other steel companies bore the entire cost of these programs. The 1960 deal marked a major victory for organized labor.[148]

The cost of this agreement and its effects on the long-term viability of Bethlehem Steel—and the entire U.S. steel industry—cannot be understated. The deal preserved Clause 2B, increased wages and improved benefits. It was not long after this agreement that imported steel began flooding the U.S. market. During the 1959 strike, many American builders and manufacturers realized that they could buy steel much more cheaply abroad than they could at home. In the tug-of-war between management and labor in the American steel industry, the unions delivered a severe blow to their opponent in 1959—one from which Bethlehem and many of its brethren would not recover.

MISSTEPS AT BETHLEHEM AND ANOTHER WAR

On July 25, 1960, Eugene Grace died at the age of eighty-three. The longtime Bethlehem leader had suffered a stroke in 1957, and his health had been declining ever since. Under Grace's leadership as president (1916–45) and chairman of the board (1945–57), Bethlehem had expanded its market reach significantly, but his patriarchal style and reliance on a vertically integrated approach would have long-term consequences for the company. In addition, Grace lacked the innovative genius and risk-taking nature that Schwab possessed. Grace was a trailblazer in expanding Bethlehem's footprint—notably in the Midwest—but he lagged well behind Schwab in keeping the company on the technological forefront of the steel industry.

Bethlehem had become increasingly insulated under Grace's watch as president, with many divisions and departments concerning themselves with only their operations. Communication suffered greatly in this segmented environment, as did creativity and an open exchange of ideas. Instead of cultivating a culture that allowed different divisions to understand how they fit into the big picture, Bethlehem became comprised of insulated units, ones in which middle managers often looked out for themselves and their immediate subordinates while keeping a watchful eye on the next opportunity for promotion. John Heinz, a speech writer for the company following the Grace era, noted:

> *The definition of intelligence or ability was to do things the Bethlehem way. And the Bethlehem way was the way we always did it in the past.... The characteristic that each department had in common was that they were fiefdoms going way back. The turf was inviolable, and prizes did not go for objective intelligence or academic training. Rarely were promotions based on merit.*[149]

Over the short term, following the strike of 1959, steelmaking remained lucrative at Bethlehem Steel. Profits in 1963 totaled $102 million, marking the twelfth time in fourteen years that the company had generated at least $100 million in revenue. Another possible steel strike loomed in the summer of 1965, but management and labor reached a last-minute agreement in which workers gained even more ground.[150]

On March 8, 1965, 3,500 American marines landed near Da Nang, South Vietnam. Their arrival marked the beginning of the U.S. ground war in this country. By the end of 1965, there were more than 200,000 American soldiers in Southeast Asia. Just as World War II had proven to be an economic boom for Bethlehem and other domestic steel manufacturers, so was the Vietnam War. Bethlehem produced small armaments for U.S. soldiers, and profits poured into the company. In 1966, Bethlehem generated $170 million in revenue, produced 21 million tons of steel and operated at around 90 percent capacity across its holdings. In Johnstown, employment held steady between 13,000 and 14,000 workers during most of the 1960s, with six blast furnaces operating at almost peak capacity. In 1965, Bethlehem reopened its coke-producing plant in the Rosedale section of the city, one that had not been in operation since 1958. The economic boom from the Vietnam War, however, only served to temporarily mask the writing on the wall for the company. The steel industry's cash spigot that had flowed for so many years across Bethlehem's holdings was about to run dry.[151]

The financial upturn during the Vietnam War was an anomaly during a transition period for the American steel industry, and Bethlehem officials failed to heed the warning signs of impending catastrophe. They continued to expand the company's geographic footprint, erecting a plant in Burns Harbor, Indiana, on Lake Michigan in 1962. This mill opened for business in 1965 and began producing sheet and plate steel for the Midwest's

The Vietnam War provided a boom in business at Bethlehem Steel during the 1960s, but it didn't change the writing on the wall for the long-term prospects of the company. *Library of Congress, Prints and Photographs Division, Washington, D.C.*

automobile industry. Burns Harbor holds the distinction of being the last vertically integrated steel mill built in the United States.

By the end of the 1960s, Bethlehem was in serious financial peril. Its net revenue dropped from $156 million in 1969 to under $100 million in 1970. On March 29, 1973, the last remaining U.S. troops pulled out of Vietnam. Three months later, Bethlehem officials announced that the company was downsizing its workforce in Johnstown from 11,800 to 7,100 over the next four years. The company also announced its intentions to decrease its payroll from $130 million to $90 million over the same four-year span.

A year later, however, in May 1974, the company announced it was reversing its decision on cutbacks in the valley. Bethlehem officials announced their intentions to build a basic oxygen furnace in Johnstown, one that would allow the company to meet the EPA's more rigid environmental standards. This decision—part of a $200 million renovation project—was strongly endorsed by Lewis Foy, who had become CEO at Bethlehem in 1974.[152]

Foy succeeded Stewart Cort, who had served as Bethlehem CEO from 1970 to 1974. Foy rose through the ranks at the company in its purchasing division. A native of Shanksville, Pennsylvania—a community roughly thirty miles south of Johnstown, near the site where U.S. Airlines Flight

Lewis Foy was born on a farm in Shanksville, Pennsylvania. He started his career at Bethlehem Steel in the purchasing department before rising to the role of chairman and CEO. *Courtesy of the National Museum of Industrial History. All rights reserved.*

93 crashed on September 11, 2001—Foy joined Bethlehem in 1936 as an accountant before serving in the army during World War II. He rejoined the company after the war. Despite repeated recommendations from other company officials to shut down the Johnstown plant altogether, the soft-spoken Foy disagreed. "We can make those plants profitable again," he told reporters.[153]

Foy was optimistic that the federal and state government—at least in the short term—would allow Bethlehem to continue producing steel through the open-hearth process, given the company's intentions to build a basic oxygen furnace. The company's CEO was also appreciative of the steps Johnstown business leaders had taken to support the company. City leaders had formed an organization called the Johnstown Area Regional Industries (JARI). In 1974, the group raised $3 million for industrial development.[154]

In 1977, however, Bethlehem reversed course in Johnstown and at several of its other holdings yet again. Foy had come to the realization that major downsizing was unavoidable. By the end of that year, 7,300 Bethlehem employees in Johnstown had lost their jobs. In Buffalo, the company laid off 10,800 members of its labor force.[155]

The blow that Bethlehem dealt Johnstown in 1977 was part of a one-two punch that the city in the valley would suffer that year. The second was delivered by Mother Nature.

5

BLOB OF PRECIPITATION

Once before, twice before, the rivers and the water had poured over us. They rose in the night and staggered us with their force. We take them for granted, and they teach us a lesson thrice. And now, on Wednesday, July 20, they rose up again.

May 29, 1889
March 17, 1936
July 20, 1977

I hope to God it never happens again.

—Larry Hudson[156]

On Saturday, July 16, 1977, a storm formed over South Dakota and began a slow progression eastward. In Iowa, it met up with an intense storm that had developed over Nebraska. When the two merged, the combined storm dropped heavy rains on parts of Indiana, Ohio and Kentucky. The southeastern edge of this tempest moved over the Johnstown region on July 19. It had become smaller in size but more powerful in its intensity when it settled over Johnstown. The National Weather Service (NWS) called it a "blob of precipitation." The NWS would later describe the amount of rainfall that it produced in Johnstown as a "once-in-every-10,000 years occurrence."[157]

From late 1976 into early 1977, heavy snowfall had accumulated across Cambria County, blanketing the mountains above Johnstown. When it melted, the region's water table rose significantly, prompting members of the U.S. Army Corps of Engineers to clear out brush and sediment deposits from channels around the county. When the snow melted and no serious flooding occurred, residents in the valley breathed a collective sigh of relief. Winter turned to spring, farmers in the surrounding mountains planted and tilled their crops and work in the mines and the mills around Johnstown proceeded as usual.

In the days before Johnstown's third major flood in July 1977, nothing particularly newsworthy was happening around the city. On Monday, July 18, coal miners returned to work after walking off the job for a week to protest cuts to their healthcare benefits. Progress was reported in talks between teachers and administrators at the Johnstown Area School District, and both sides were optimistic that a strike could be avoided.[158]

It wasn't a Charlie Chaplin film showing in downtown Johnstown this time around; it was a Paul Newman movie with a distinct taste of the city. *Slap Shot*, a comedy about a minor league hockey team in a Northeastern mill town, was running at the city's Act One Theater on the evening of Tuesday, July 19. In the movie, the fictitious "Charlestown Chiefs" resort to dirty, thuggish play on the ice in order to compete against their more talented opponents. The movie was actually filmed in Johnstown and enjoyed financial success, generating more than $18 million at the box office. *Slap Shot* eventually drew a cult following that continues to this day; it is considered by many to be the best hockey movie ever made. Nancy Dowd wrote the screenplay for the film, one based on the experiences of her younger brother, Ned Dowd, who played minor league hockey during the 1970s. One of Ned Dowd's stops during his minor league hockey career was Johnstown. He played for the Johnstown Jets, an affiliate of the St. Louis Blues, from 1973 to 1975.[159]

In baseball-related news, on July 19, the top players in the American and National Leagues were preparing to face one another in the Major League All-Star Game at Yankee Stadium. Many Johnstowners tuned their television sets to NBC that evening to see how Pittsburgh Pirates right fielder Dave Parker would fare in the mid-summer classic (he singled and scored a run in the National League's 9–8 victory).

"I was down in the funeral home in the second-floor apartment with my father, watching the game," said Bill Hindman, director of Hindman Funeral Homes in the Morrellville section of the city. "We realized that it was raining excessively hard, but we'd all seen rainstorms before. Like everyone else, we didn't realize that we were going to be getting in excess of nine inches that night and how it was going to turn our lives around."[160]

Despite a forecast that indicated the likelihood of precipitation at around 30 percent, it began raining heavily in Johnstown around 6:30 p.m. on July 19. An hour later, flash flood warnings were issued for Cambria and six other western Pennsylvania counties. Blinding lightning accompanied the heavy rains throughout much of the region. "The lightning was unbelievable," said Janet Smetto-Mical, who lived along Power Street in Cambria City. "Scary and loud." Lorain Borough volunteer assistant fire chief Rich Horner began rescuing people from low-lying areas hours after the rain started. "We were going around to houses and getting people out," said Horner.[161]

Johnstown mayor Herb Pfuhl took a call from a member of the city's police department early in the evening. Born in Johnstown in 1928, Pfuhl was the son of German immigrants who had moved to the valley in the early 1920s. His mother, Anna Schweitzer, was a relative of legendary doctor and theologian Albert Schweitzer. Pfuhl graduated from Johnstown Area High School in 1946 and enrolled at the University of Pittsburgh. After serving in the navy during World War II, he became a teacher in the Ferndale-Dale School District in the southern section of the city. A Republican in a union town, the straight-talking Johnstowner from the city's West End was elected to his first term as mayor in 1971, after serving as Cambria County treasurer for four years.[162]

The police officer who called Pfuhl told the mayor that the flash flooding was creating significant problems around the city. Pfuhl and his son, along with city employees Dominick Genovese and John Sprincz, drove around town to assess the situation. Around midnight, the group visited the Meadowvale School in the Hornerstown section of the city. The school had become an emergency shelter location. When the mayor and his entourage arrived, they discovered that around one hundred people had taken shelter in the building. Those who had gathered at the school moved to the second floor around 1:00 a.m., after water damaged a rear wall of the building. During the night, several members of the group removed the lane-marker ropes from the school's pool and used them as lifelines to save people trapped in nearby homes.[163]

Floodwaters rose quickly in many neighborhoods as the downpour hammered the region for seven straight hours. By 11:00 p.m., at least four and a half inches of rain had fallen on much of the city. "The storms never quit but increased in intensity," wrote Larry Hudson in the *Tribune-Democrat*.

A bridge near Walnut and Washington Streets was consumed by the floodwater. *Johnstown Area Heritage Association.*

"By 2:00 to 3:00 a.m. Wednesday, thousands were aware that the weather was going berserk." In one forty-minute period—between 2:50 and 3:30 a.m.—more than two inches of water fell on the city.[164]

Around midnight, a police spokesman announced over the radio that the city was closed to all incoming traffic. Few received the message, however, because electricity was knocked out across much of the region. Johnstown's police and fire radio networks were not functioning either, and many phone lines were down. "We've got full crews out," said a telephone repair worker, "but right now, we're just hitting and missing." At 2:00 a.m., Pfuhl declared the city a major disaster area. "The devastation is unbelievable," he said. The mayor established a command post at the WJAC Building in Upper Yoder Township.[165]

Stepping into the communication void to keep the community at least somewhat connected were people with CB transmitters. The voices of CB-ers crackled all evening and helped more than a handful of stranded Johnstowners. "Breaker one-nine, what's the situation downtown?" one CB-er asked another. "We need a boat here to take two doctors into Lee

Behind one of Bethlehem Steel's shops, floodwater rose above the Washington Street Bridge. *Johnstown Area Heritage Association.*

Hospital....Can someone please get a landline to Harrisburg? We've got people trapped in West Taylor Township and we need a helicopter," said another. "Come back Ferndale, we've got a boat here....A hitch, we need a hitch in front of Mercy Hospital to move a boat."[166]

Ron Gatehouse, the night-side editor at the *Tribune-Democrat*, called the newspaper's managing editor George Fattman around 3:00 a.m. He told his boss that the water on Locust Street outside the paper's offices was up over the mailbox. "When I got off the phone, I called [publisher] Dick Mayer," said Fattman. "He said, 'That can't be. I don't believe it.' And he hung up. Then he went over to the top of the incline and he called back and said, 'It is flooded!'"[167]

As was the case in 1889 and 1936, the rain that hit the mountains above the city poured into the rivers, creeks and streams. This water continued into the valley and combined with the rainwater pummeling the city. The storm—which had planted itself over the southern sections of Cambria and neighboring Indiana County—did not budge. National Weather Service investigator Dave Sisk said that the storm's unusual lack of movement or

The floodwater wasted no time washing away earth behind the Solomon Run Homes before pounding the building itself, carrying away large chunks of these apartment units. *Johnstown Area Heritage Association.*

dissipation was "an act of God." These were words that generations of Johnstowners had heard before.[168]

The Widman Street exit ramp was washed away, as were a number of homes on Solomon Street. Eighteen residents of the Solomon Housing Project, located in one of the hardest-hit areas of the city, lost their lives. Downtown, water flooded the basements and ground floors of numerous buildings, including Lee Hospital and the *Tribune-Democrat* offices, interrupting electricity, gas and telephone services. The first floor of Lee Hospital housed the cafeteria and served as the hub for the facility's electrical system. Officials at the hospital instituted emergency procedures when the facility lost electricity. Over the next four days, 227 Lee patients were moved to either Mercy or Conemaugh Memorial Hospitals. Some of these patients were transported by boat from Lee to Conemaugh. From July 20 to July 23, Lee admitted no patients. Power outages also left staff members at the city's Red Cross Building scrambling. Without power, the relief agency lost use of refrigeration units that kept its blood supply cold. Helicopters soon arrived at the Red Cross, and the blood was transported to the Galliker's Dairy Warehouse in the city, where it was placed back under refrigeration.[169]

Just as they had in 1889 and 1936, Red Cross staffers and volunteers worked tirelessly in the summer of 1977 to help the city recover from another catastrophic flood. *Johnstown Area Heritage Association.*

Two major fires ignited during the flooding and burned throughout the night in Hornerstown. The Robel Construction Company was destroyed by flames as floodwaters raged, and another fire started a few blocks from this site. Rising water prevented firefighters from getting their equipment close enough to effectively battle the blazes. Twelve bodies were discovered near the Cambria Chair Company in Hornerstown early Friday morning.[170]

Wayne Cook; his wife, Edna; and their four children lived in a small, redbrick home that the couple had purchased during the late 1950s near Lambs Bridge. Around 1:00 a.m., Wayne Cook was alerted to the seriousness of the situation. "My son was sleeping on the parlor floor and came upstairs at 1 o'clock and said, 'Dad, we're getting water in the parlor.' When that bridge up there let go, boulders the size of cars came down and hit my house." The family watched the floodwater roar past their home. Cook reached his hand out the window several times and felt the water rushing by.

"A big tree came through the dining room window, and it knocked our kitchen wall out," said Edna Cook, who served as head nurse in the children's ward at Conemaugh Memorial Hospital. In the days that followed, she administered tetanus shots to hundreds of area residents. Wayne Cook was humbled by the generosity demonstrated to his family following the destruction. "People around here, you just don't find this kind anymore," he said. "They must all be givers in this part of the country."[171]

LAUREL RUN DAM COLLAPSES

The loss of life and property in Tanneryville, a community of three hundred residents located in West Taylor Township, was catastrophic. The original Laurel Run Dam was constructed above Tanneryville in 1869 to help Johnstown meet its water needs, but it was abandoned in 1910. Work on a new dam—the Laurel Run Reservoir No. 2—began in 1915 and was completed four years later. The dam measured forty-two feet high at its spillway, stretched 620 feet from end to end and was large enough to contain 101 million gallons of water. In 1943, engineers noted that the dam was inadequate to handle a "large storm event." A 1959 state-required assessment concluded that the dam's spillway was less than half the size expected under Pennsylvania guidelines. Despite these concerns, no action was taken to increase the spillway's capacity. In 1970, the dam was classified as "high hazard" under criteria established by the Federal Emergency Management Agency. High-hazard dams are defined as those where "mis-operation or failure will probably cause loss of human life."[172]

Around 2:30 a.m. on July 20, 1977, the Laurel Run Dam gave way. Hours earlier, water had begun pouring over the breast of the dam because the spillway was not big enough to contain the water that had already accumulated. By the early hours of July 20, twelve inches of rain had fallen on the Laurel Run Basin in a matter of hours. The collecting water—more

The Laurel Run Dam was built to contain up to 101 million gallons of water, but after Mother Nature dumped almost twelve inches of rain on the Tanneryville area on July 20, 1977, its earthen wall collapsed. *Johnstown Area Heritage Association.*

than 100 million gallons—could not pass through the dam's spillway fast enough, and it began to pour over the earthen wall's breast. As the water rushed over the top of the dam, a notch formed, and the breast began eroding. The dam wall became more and more compromised. Eventually, the water pushed through what remained of the breast, sending a deluge roaring toward Tanneryville. No one was present at the dam site to provide a warning to the people below. "The water came up over the earthen breast of the dam," said Ed Cernic, an area businessman who would help organize the Tanneryville Flood Recovery Association and the Tri-County Flood Recovery Group. "There was no concrete there; it was just dirt. The water was so fast and so devastating." Tanneryville resident Vickie Daniels clung to her refrigerator to avoid being pulled into the water. "It was almost black, like a river of debris," she said. "I remember watching it carry a firetruck like it was nothing." Daniels and her dog were pulled to safety with an extension cord.[173]

Thirty-four people were killed in Tanneryville, and hundreds more were injured. An estimated fifty to sixty people were swept up into the raging floodwaters after the dam broke. Carol Burns, who lived on Cooper Avenue, lost her husband, daughter and mother-in-law. Burns and her mother were playing bingo at Roseland Hall on Fairfield Avenue on Tuesday night. "On our way home from bingo, we made it almost to Cernic's Gas Station," she said. "There were people there, and they wouldn't let us go any farther than that. They picked my mom and I up and then they took us up behind Cernic's Gas Station. We stayed with a family that was taking all of these people in. We spent the night there."[174]

On Friday morning, Burns learned the fate of her family members. "They told me that they found my husband and my mother-in-law down at the wire mill, in that area, and they found my daughter down in Seward, underneath the bridge where Routes 711 and 56 come together. It took quite a while to find out where my husband and daughter were. All these people were there, but nobody would really tell you anything." There was nothing left of her family's home on Cooper Avenue, so Burns stayed with her parents for several months before moving into a mobile home, one of many brought to the valley by the Department of Housing and Urban Development. Twenty-eight homes were destroyed by the flooding in Tanneryville, and many others had to be torn down in the months that followed.[175]

The disintegration of the Laurel Run Dam caused the water that it had held back to hit the Conemaugh River with such force that water began flowing upstream. This backflow caused flood levels within Johnstown to rise quickly. The backwash carried trees, cars, rocks, gravel and parts of residents' homes into downtown Johnstown, exacerbating the flooding problems there. Debris temporarily blocked the floodwater and prevented it from flowing downstream. "I was sitting on my front porch steps," said Cernic, who lived in the Cooperstown section of the city, near the spot where Laurel Run enters the Conemaugh River. "It was about 3:30 a.m. I was looking across the river and saw a train backing up and the water rising. When I saw how it was rising, I ran back into the house, and I told the family to go for higher ground." Cernic estimated that the Conemaugh River was four feet below the top of its channel when the breast of the Laurel Run Dam washed away. He said that within five to ten minutes of the dam's failure, the river rose eleven or twelve feet before escaping its channel.[176]

William Keiper also lived in Cooperstown. He saw a man become trapped in his car after the Laurel Run Dam gave way and he went to try and help the motorist. Keiper said he saw the river current moving

"There ain't no more Tanneryville," said a national guardsman after surveying the damage to this community on the western edge of Johnstown. *Johnstown Area Heritage Association.*

upstream as he assisted the man. Other residents of Cooperstown said they saw a crest of water moving upstream on top of the Conemaugh River current, which they said was headed downstream. "The debris and flow that swept into the Conemaugh River as a result of the Laurel Run Dam failure caused the Conemaugh to rise fifteen feet higher than it would otherwise have done," said David Wojick, a civil engineer who prepared a report examining the causal factors of the flood. "Without this extra rise, the river would not have left its channel and the city of Johnstown would not have flooded."[177]

Other investigators disagreed with Wojick's conclusion, arguing that the city would still have experienced major flooding had the Laurel Run Dam not failed. The general consensus was that serious flooding was inevitable because of the amount of rainfall the region received in such a short period of time. Eugene Armicida, a hydrologist with the Army Corps of Engineers, believed that the dam's collapse had no effect on the water that backed up in downtown Johnstown during the early hours of July 20. Another government engineer told reporters that there was "very little factual data in [Wojick's] report that supports his argument."[178]

"It is unbelievable what happened when the Laurel Run Dam burst," said city resident Jack Biagis. Following the dam's break, sections of Tanneryville were accessible only by helicopter until emergency crews established paths for four-wheel-drive vehicles. This isolation in the immediate hours following the dam's collapse contributed to the death toll. And to make matters worse, in the months prior to the flood, West Taylor Township's supervisors had voted against participation in a federal flood insurance program. Much of the rebuilding costs would fall directly on home and business owners. "After the horse is stolen, then you lock the barn," said Cernic.[179]

The Laurel Run Dam was never rebuilt, and the site is now part of Laurel Run State Park. "We investigated this dam in the '60s," said Elio D'Appolonia, a geotechnical engineer who returned to the city for another investigation following the 1977 flood. "Its deficiencies were recognized and reports prepared for modification, but for various reasons, over a period of one-and-a-half decades, remedial steps or new construction was not taken. If the dam had been upgraded in accordance with today's prudent engineering practice, it would have been able to store and/or pass the storm."[180]

THE AFTERMATH

The rain stopped falling around 4:00 a.m. on Wednesday, July 20. The storyline was similar, yet different, from the floods in 1889 and 1936. The primary cause of the 1889 flood was the failure of the South Fork Dam. The 1936 flood was the result of a large system of heavy rains that extended across much of the Northeast and Mid-Atlantic regions of the United States. The 1977 Johnstown flood was the result of unusual, extremely powerful thunderstorms, coupled with the collapse of the Laurel Run Dam. "The specter of the Great Flood of 1889 and the St. Patrick's Day flood of 1936," read an editorial in the *Tribune-Democrat*, "hovered as heavily over the city as did the fog that hid, from time to time, the evidence of Wednesday's flooding."[181]

After surveying the damage, George Single of the Army Corps of Engineers concluded that the flooding was worse than that of 1936. Single and other members of the corps inspected around thirty dams throughout the area in the weeks following the flooding. The group reported that the Laurel Run, Sandy Run and Peggy Run Dams—all located on tributaries

Many roads, including this section of Ohio Street, were not traversable following the flood due to cracked and displaced pavement. *Johnstown Area Heritage Association.*

of the Conemaugh River—had failed. The other dams in the area were described as stable, but many of them were filled to capacity, and water splashed over their spillways. Where possible, corps members released water from these dams.[182]

What Johnstowners woke up to on July 20 was a mess unlike any most of them had ever seen. Mud and murky standing water was everywhere. It was in basements and yards and on sidewalks and roadways. Submerged cars outnumbered vehicles on dry land in some parts of the city. Downed tree branches and cracked pavement made driving an impossible proposition in many places. National Weather Service investigators Dave Sisk and Don Wilson reported that 8.9 inches of rain had fallen on parts of the city between 6:30 p.m. on Tuesday and 4:00 a.m. on Wednesday.[183]

There was no telling what you saw or where you saw it—propane gas tanks, kitchen cabinets, garbage cans, refrigerators, dog houses, clothes and strips of wood floated in murky water or rested in mired muck. A car leaned against a telephone pole downtown, and a washing machine rested on top of the vehicle. Business owners assessed the damage, walking barefoot through their stores and restaurants, carrying their socks and shoes in their hands. More than twenty railroad cars had been swept up by the raging Conemaugh

River, as had a number of mobile homes. One constant throughout much of the region was the foul odor of raw sewage. It not only stung residents' noses, but it also burned their eyes.[184]

The storm had dealt different fates to different parts of the region. Many communities around Cambria County were hammered, while others were spared. Parts of the Laurel Run Basin had received almost twelve inches of rain in eight hours. Just a few miles from the storm's bullseye, however, the rain was heavy but short-lived. One-half inch of rain fell on Boswell, a community only a few miles south of Johnstown. To the west, two inches of rain fell on New Florence. To the north, four and a half inches of rain fell on the county seat of Ebensburg.[185]

Twenty-two bridges in the region suffered serious damage, and more than 150 miles of roadway around the city and surrounding communities were also damaged. The Consolidated Rail Corporation (Conrail) reported that thirty miles of railroad tracks from South Fork to Bolivar—a community twenty miles northwest of Johnstown—were either washed out or blocked by debris. The South Fork branch of the railroad and the St. Michael railyard were both washed out by floodwater. Ten of Conrail's locomotives were under water, and another ten coal hopper cars were carried into the Conemaugh River.[186]

Amtrak halted service to Johnstown temporarily because of the flood damage. Eastbound passenger trains were stopped in Pittsburgh, and westbound ones were only permitted to travel as far as Harrisburg. Passengers were then bussed from Pittsburgh to Harrisburg and vice versa. When Amtrak resumed service through Johnstown, trains were required to move at a speed of no more than thirty miles per hour in a twenty-two-mile radius around the city.[187]

The city's inclined plane served as a rescue mechanism again, lifting more than 1,500 people to higher ground. The incline was shut down because of a power outage around midnight on Tuesday, during the height of the storm, but it was soon back up and operating. On Wednesday and Thursday, residents rode the incline back down into the city, returning to their homes and businesses to assess the damage. Downtown, the Swank Building—located at the intersection of Main and Bedford Streets—became a temporary city hall.[188]

Congressman Jack Murtha, who represented Johnstown and the surrounding area in the U.S. House of Representatives from 1974 to 2010, told reporters on Wednesday that the flood had displaced roughly seven thousand people. He noted that more than one thousand residents had sought refuge at "care centers" throughout the city.

Murtha was in Washington, D.C., speaking to a group of soldiers at a marine corps base when he was made aware of the flooding on Wednesday morning. "They came in and slipped him a note that said 'Johnstown flooded,'" said his wife, Joyce Murtha. "His first reaction was, 'We get floods all the time,' but then he thought, 'I better check this out.' He called and found out that it really was a devastating flood." Jack Murtha then called the White House and requested that a presidential helicopter pick him up so that he could get back to Johnstown quickly, and one was made available to him. Upon arriving in the valley, the congressman coordinated recovery efforts from this helicopter, which rested on a landing pad at the city's airport.[189]

Jack Murtha was the first Vietnam War veteran elected to the U.S. House of Representatives. He served in Southeast Asia from 1966–67 and was elected to Congress in 1974. *Wikimedia Commons.*

On Wednesday evening, around 275 people sought refuge at the Richland Junior High School. The following evening, the number of people gathered at the school grew to around 450. Hundreds of cots were flown into the valley on national guard helicopters from a barracks in Ebensburg and delivered to the school. Pennsylvania governor Milton Shapp, who visited the school on Thursday, instructed Richland School District assistant superintendent Edward Pruchnic to purchase anything that he felt was needed at the shelter. The University of Pittsburgh's Johnstown campus in Richland Township, Saint Francis College in Loretto and Mount Aloysius College in Cresson also opened their dormitory rooms to displaced residents.[190]

Many who arrived at the shelters were shell-shocked. Peggy Young fled her Solomon Homes apartment with her two children and sought refuge at the Richland Junior High. "I waded through water and climbed a mountain of mud," she said. "When I got to the top, I thought I'd meet God. This might sound crazy, but when I took my shower in the locker room here, I saw floods and boulders coming down on me."[191]

The bodies of those who lost their lives at the Solomon Homes began to create a foul odor in the hot and humid summer air. "Our best method here is to smell and snoop," said Andy Bowser, who led a group of volunteer firemen searching for decomposed bodies in the housing project. "All Tuesday night, I was upstairs on the third floor of my building," said Solomon Homes tenant

Charles Trail. "I thought we wouldn't make it. The first thing I had on my mind was to kill time when the storm started. My little daughter is afraid of lightning, so I started to strum a little on the mandolin—anything to take her mind off it. I looked outside, and the water was rising. We left, hanging onto a rope, walking in chest-deep water. I never thought we'd make it."[192]

Murtha and Pfuhl surveyed the damage by helicopter on Wednesday, along with other state and federal officials, including Governor Shapp, Lieutenant Governor Ernest Kline, U.S. senator John Heinz, Secretary of Health Leonard Bachman, Presidential Personal Assistant Greg Schneiders and former Pennsylvania Department of Transportation secretary Jacob Kassab. Shapp had called Kassab out of retirement to serve as his personal representative to the area. Schneiders reported back to President Jimmy Carter that the flooding was "a disaster of major proportions." James Hogan, bishop of the Roman-Catholic Church's Altoona-Johnstown Diocese, also visited to assess the damage. He ordered special collections conducted for flood survivors at all of the Catholic churches in the eight-county diocese.[193]

Pfuhl estimated the damage to the city to be somewhere around $100 million. Shapp disagreed with the mayor's assessment, suggesting that $200 million was a more realistic figure. On Wednesday night, a special meeting was held at the White House to discuss the damage and organize relief efforts. Heinz and fellow Pennsylvania senator Richard Schweiker were in attendance, as was Stuart Eizenstat, a special assistant to President Carter for domestic affairs. Also participating in the meeting was Thomas Dunn, director of the Federal Disaster Assistance Administration. During the seventy-five-minute meeting, the group viewed video footage of the flood damage. "There is no doubt in my mind that the president will formally declare the seven counties a disaster area tomorrow," noted Heinz following the meeting.[194]

On Thursday morning, President Carter signed a disaster relief bill that provided Johnstown with federal funds for temporary housing, assistance with repairs, home, business and farm loans, unemployment compensation and emergency food stamps. The number of wage earners put out of work in the weeks immediately following the flooding was estimated at more than twenty thousand by the Bureau of Employment Security. In Harrisburg, the Pennsylvania House and Senate took steps to help Johnstowners as well, voting unanimously to increase the state's disaster relief fund from $1 million to $5 million. Ed Cernic made a trip to the White House, meeting with President Carter to lobby for additional low-interest loans for small businesses and homeowners.[195]

It was difficult for officials to estimate the extent of the damage following the 1977 flood, because, among other reasons, you couldn't tell what was at the bottom of many of the countless piles of debris. *Johnstown Area Heritage Association.*

Rescue squads from Harrisburg and Kittanning arrived in the valley on Saturday to assist Johnstown's fire and police forces with the task of searching for bodies and removing them from the sea of debris left in the flood's wake. "Until the heavy debris is moved, and the mud and sludge as well, we won't know who's under there," said John Burkett, Johnstown's deputy fire chief. In many of the hardest-hit sections of the city, the only access to them in the immediate days following the deluge was on foot. "Where Solomon Road was, a river is now," said Burkett. "That area is so badly torn up that we're actually going to have to build a road up into there before we can get heavy equipment in."[196]

The mining town of Windber to the east of Johnstown sustained major damage. Eight inches of rain forced Paint Creek and Seese and Elton Runs out of their banks, and Weaver Creek spilled out of its channel. The bodies of Andrew Koharchik and his wife, Margie, were discovered on the grounds of Windber's Slovak Club when the floodwater receded on Wednesday. The couple had fled the basement bar that they managed, but Margie was concerned that some of their business's financial assets might be damaged. She went back into the building to retrieve them. Her husband followed, and both were killed when floodwater enveloped them.[197]

Cars and trucks from Shaffer and Baumgardner Motors on Windber's Jefferson Avenue were swept to various locations around this mining town. A section of Route 56 near the Scalp Level Bridge suffered serious damage. Windber mayor Thomas Panetti instituted a 9:00 p.m. curfew in the weeks following the flood in an effort to limit looting. He estimated property losses in his community at around $30 million. "I thought it was the end of the world," said Windber police chief Melvin Causer.[198]

"It was raining hard all afternoon," said Windber resident Bob Carnville. "There was a gentleman at the bar who said he saw a green sunset. It was like an omen. My little sister woke me up at 12:30 a.m. Wednesday morning and said the water was rising in the basement. The water just kept rising for an hour. We were scared to death. In the house, we just prayed. The whole family was scared—except my dad. He was making plans for rebuilding while the water was rising." Sixty children from Camp Harmony in nearby Hooversville formed a bucket brigade the day after the flood, cleaning up mud and bailing out water from Windber residents' basements.[199]

The Red Cross set up care and food centers at a number of sites in Windber, including the Tenth Street Russian Club, the Scalp Level Fire Hall and VFW Building, the Church of the Brethren in Tire Hill, St. Mary's Greek Orthodox Church, Conemaugh Township High School and Windber Area High School, where 1,500 meals were prepared each day in the week following the flood. "We're advising the people not to drink the water or to let their children drink it until it's been boiled," said Mayor Panetti. "We are also urging parents to prevent their children from playing in the mud from the storm. There's raw sewage in that mud."[200]

In Seward, a community to the west of Johnstown in Westmoreland County, fifteen bodies were recovered. Seven of them were found in the Hoover Trailer Park. The cloverleaf where State Routes 53 and 869 meet in Liberty Park was rendered impassable. Used cars at Elmo's Service Station were stacked on top of each other in a monochromatic scene.[201]

North of the city, in the communities of Summerhill and Sidman—and not far from the site of the infamous South Fork Dam—damage from the flood was estimated to cost $2.5 million. Water had breached the Highland Sewage and Water Authority Dam above Sidman. There was no one at the dam site when its thirty-foot-tall earthen breast yielded to the weight of the water. "All of a sudden, I just looked up, and the water started rising," said Don Weaver, who lived several miles below this dam. "It was coming at sort of an angle. I ran upstairs and grabbed my wife and we started running. If I'd been ten or fifteen minutes later, I'd be dead." In South Fork, the

Pennsylvania National Guard members assisted residents in many ways, including distributing canned and bottled water. Ironically, many residents did not have water following the 1977 flood because pipelines had been damaged. Others were told not to drink the water from their faucets for fear of contamination. *Johnstown Area Heritage Association.*

water rose nine feet in the span of several minutes, and more than fifty families evacuated their homes. Marvin Mervine tried to escape but was unsuccessful. As he backed his car out of his family's driveway, the bank supporting it collapsed, killing the Summerhill native.[202]

Twenty miles northeast of Johnstown, in the village of Portage, the Little Conemaugh River and Trout Run poured out of their banks, forcing a number of houses off their foundations in the borough's lower end. Roughly seventy-five families were forced to flee their homes. Portage mayor Thomas Mitchell estimated the damage in his town at a cost of around $1.5 million. The flood destroyed five bridges in this community on Sonman, Conemaugh and Jefferson Avenues and Caldwell and Main Streets. During the flooding, seven Portage firemen became trapped while they tried to evacuate residents. As the firefighters were returning to their truck, the line that they were holding on to broke. They were forced to hang on to a fence until the water subsided.[203]

The 1977 flood killed eighty-six people. Many of these victims lost their lives because the Laurel Run Dam failed. Residents had gone to bed on the evening of July 19 unaware of the seriousness of the situation. The remains of victims were found half submerged in streams, buried in garbage piles and washed up on rocks. Four days after the flood, several Blairsville residents discovered four bodies near the Conemaugh River Flood Control Dam in this Indiana County community, thirty miles downstream from Johnstown.[204]

The nature of the 1977 flood created serious problems. "It was a sudden flood and a very high-velocity flood," said Sidney Goldblatt, lead pathologist and lab director at Conemaugh Memorial Hospital. Goldblatt performed more than seventy autopsies on victims in the week following the deluge. "Some bodies were impaled by trees and wrapped in wire. It was clear that many of the bodies experienced trauma."[205]

Funeral home director Bill Hindman coordinated arrangements for the last discovered victim, a girl whose body was recovered in May 1978—ten months after the flood. "It was a little Gibson girl from out in Tanneryville. She was nine years old," said Hindman. "Her remains were found on Gooseneck Island downriver. She was identified by the clothing."[206]

"To be alive at dawn on Wednesday was equivalent to being given—for no reason, with no justification—the second chance no one gets," wrote the *Tribune-Democrat's* Nina Kalinyak.

> *The opportunity to reassess and do it right this time. And there is something of the guilt the survivors of concentration camps speak of when they ponder about why they are alive and others dead. Surely, they reason, I did not survive because of merit of spirit, or deed, or ability. And neither did we. The threat of death appears to drive us to declare ourselves for life.[207]*

James Smith and his two sons, Troy (eight) and Todd (seven), were among those who did not survive. After water broke through a creek wall near Messenger Street downtown, the released torrent crashed into the Smiths' home and buckled the structure. "It jumped the channel at Bantley Hardware and went straight down Messenger Street," said Norm Verhovsek, who lived on Highland Park Road. "It went right down Messenger Street, crossed the railroad tracks and right down to the river."[208]

The Smith family shared a duplex with William Emmell and his wife. The Smiths and Emmells kept in contact with each other throughout the night by knocking on a wall between the two units. Shortly after midnight, James Smith decided that he needed to get his family out of the house. They were about to leave the duplex around 2:00 a.m. when it began to break apart. Mrs. Smith was holding onto the back of her son Todd's shirt when a wave of water caused her to lose her grip. Shortly thereafter, her husband and their oldest son, Troy, were swept into the swirling waters. "We watched the Smith house go and then a four-car garage behind it went, too," said neighbor Ricky Pfeil, who was ten years old at the time. He had played outside with Troy and Todd earlier that evening. "We thought we were next," said Pfeil.[209]

Pfeil's parents decided they needed to leave their house and that they should try to swim to his grandparents' home, which was just across their yard. A wall of water, however, prevented them from opening their front door. Pfeil lost his grandfather in the flood. He was swept away while trying to save his dog. "The water just took him," said Pfeil. "It was probably a week later when they found his body in a pile of debris down by the old chair factory."[210]

Randy Teeter and his parents fled their home in West Taylor Township after a stream became a twelve-foot wall of water. A tree fell on Teeter's father, but he was able to get out from under it when raging water carried it away. Teeter's mother, Dorothy Ann, was swept up into the water and drowned. "I went flying into the water," said Randy Teeter. "I grabbed hold of some debris and was swept along in the darkness. Then I stopped floating. Other debris piled up against my legs, and I was pinned." Teeter was eventually able to lift the debris off his legs three hours after he was pinned down. He climbed up the side of a bank and out of the floodwater.[211]

Daise Heslop survived all three of Johnstown's major floods. The youngest of Jobe and Eleanor Morgan's five daughters, Heslop, was born on May 6, 1883. She was six years old at the time of the Great Flood of 1889. Her father and grandmother were killed in this deluge. "I guess I was too young to be scared," said Heslop. "My father called us over to the window to show us the water rising. He told all us kids to get on the third floor, but a log caught his arm, and he drowned." Several men carried Heslop and her sisters out of their home in May 1889 and over rooftops to safety.[212]

During the 1936 flood, Heslop became separated from her husband for a day before the two found one another during the cleanup. She was on jury duty in the Moxham section of the city when the floodwaters rose in March 1936. On the evening of July 19, 1977, Heslop watched the rainfall from the window of her apartment at the Allegheny Lutheran Home in Westmont. "I love Johnstown," said the ninety-four-year-old after seeing her hometown flood for a third time. "I wouldn't live any place else."[213]

It was impossible to initially identify many of the bodies that were recovered following the 1977 flood. Cambria County coroner Joseph Govekar assigned letters to victims who could not be identified, along with some descriptive information:

> *Body C: White male, between 30 and 40, well-developed, pronounced dead at 11:50 a.m.;*
> *Body D: White male, mid-50's, pronounced dead at 4:30 p.m.;*
> *Body E: White male, late 50's, pronounced dead at 4:30 p.m.*[214]

An emergency morgue was established at the East Hills Elementary School in Richland Township. Bodies were placed in the boys' locker room at the school, which was below ground level, in an effort to keep them cool. They were initially packed with ice before the school ran out of it. Officials contacted the Galliker's Dairy Company, which delivered more ice to preserve the bodies in the 80-plus degree Fahrenheit temperatures. A cooler at Schrader Florists and Greenhouses in Geistown was also utilized as a temporary storage location for bodies. Victims' family members reported to the school and the greenhouse to identify the remains of their loved ones. The bodies were then moved to various locations for embalming.[215]

By late Wednesday afternoon, the national guard had established an emergency morgue in a parking lot next to Richland High School. The guard asked for volunteers from funeral home staffs around the region. "They were gathering up anybody who could embalm," said Hindman. "There was this little isolated area out on Central Avenue that included the Henderson Funeral Home. They had electricity, and they had water, so we were able to embalm over there for several days. When I walked into the prep room, there were two little brothers. Their house had collapsed on them. They were the same age and had the same hair color as my son."[216]

The summer heat complicated the challenges for volunteers embalming the victims. "It was hot," said Gary Henderson, director of Henderson Funeral Homes. "We did whatever identification we could, but we didn't know who anybody was. So, you marked down 'this color of ring' or 'this color of pants.' We jotted everything down and put a tag on them."[217]

Amid all of the sadness and grief, children at East Hills Elementary School played tag and tossed rubber balls with each other. "You meet an awful lot of good people at a time like this," said Cambria County's deputy coroner Art Keiper, whose mother had survived the Great Flood of 1889. "Sometimes, death can bring out the best."[218]

The bulldozers used to clean up debris following the flood were dwarfed in many instances by the massive piles of tree limbs, brush, mud, cars and parts of homes and businesses that they pushed into even larger piles. *Johnstown Area Heritage Association.*

At East Hills Elementary and other emergency shelters, survivors posted messages on bulletin boards in a manner similar to the way families posted messages in search of loved ones in New York following the terrorist attacks in 2001. "Anyone who knows the whereabouts of Connie, Maria and Roxanne Hellman, please call Ron Hellman," read one note at the national guard armory near the Johnstown Airport. The Salvation Army served sixty thousand meals to people at shelters during the week following the flood. Housing repair assistance, farm loans, unemployment compensation and income tax counseling services were also provided at many of the shelters.[219]

Bulldozers pushed massive piles of mud and debris off the streets. Eventually, much of this mud turned to dust as it baked under the hot sun. Soon, residents had dust storms accompanying the stench of raw sewage and "flood mud" in their communities. Flood mud was the name given to mud that had mixed with oil and acid from the mills. National guard helicopters hovered over the city in the early daylight hours on Wednesday, rescuing residents from rooftops. "We are going after the living first and then the dead will come later," said Richland Township police chief James Mock.[220]

"I pulled five bodies out from all over the city," said marine corps gunnery sergeant Harvey Freville. National guard captain Mario Meola flew his Chinook chopper into Seward. More than one hundred people were rescued by helicopter and boat in this community alone. "Some were so scared, they didn't want to get into the aircraft," said Meola, a thirty-year-old Vietnam War veteran who flew combat missions in southeast Asia. "We had to order them to get in. Some were suffering from exposure; some had lacerations and back problems. One woman had a ruptured spleen." The Richland Fire Hall became a field hospital, and the nearby parking lot at the Mason Department Store was used as a helicopter landing pad. In addition to their rescue missions, helicopter pilots also delivered fuel to shelters and the city's hospitals to power emergency generators. Most of the helicopter missions were halted early Wednesday evening, when exceptionally dark skies made them too dangerous to continue.[221]

Similar to the days and weeks following the 1889 and 1936 floods, police and military support was not lacking. Those who provided services to Johnstown in July 1977 included the 103rd Armored Battalion and 876th Engineering Battalion from Johnstown (29 officers and 343 enlisted men); the 107th Field Artillery from Pittsburgh (23 officers and 203 enlisted men); the 108th Field Artillery from Carlisle (21 officers and 252 enlisted men); and the 28th Military Police from Johnstown and Altoona (10 officers and 168 enlisted men). Other assistance was provided by a helicopter unit from Indiantown Gap.[222]

"The biggest help in the recovery was the volunteers," said Lorain Borough volunteer assistant fire chief Rich Horner. "You had all these people from different organizations coming in, bringing supplies. We had a woman in Lorain Borough, she was cooking food in her kitchen for everybody who was working."[223]

Plenty of donations also poured into the valley: 1,100 cases of baby food from the H.J. Heinz Company; 3 tons of ice from Frazier Refrigeration; blankets and bedding from Sears; 2 tractor-trailer loads of water from Canada Dry. The generosity was so great that it proved to be problematic. "Tons of goods are arriving, unplanned and unorganized. There is no place to put it," said John Comey, a public relations officer for Cambria County's State Council on Civil Defense. Comey urged residents to organize their donations through the Red Cross or Salvation Army.[224]

More than 15,000 people lost electricity during the flood, and phone service was also down throughout much of the city. General Telephone sent 250 employees from its Erie and York offices to assist in service restoration.

WJAC-TV was off the air for thirty-five hours after its transmitter on Laurel Mountain was damaged. "There was no radio service and no TV service and after a couple of hours, there was no water service because the water took out the pump station in Dale," said Norm Verhovsek. The *Tribune-Democrat* temporarily moved its operations to WJAC's headquarters in Upper Yoder Township after it lost power. In the days following the flood, the *Tribune-Democrat* was printed by one of its competitors, the *Tribune-Review* newspaper in Greensburg, a community an hour west of Johnstown.[225]

Even after electricity was restored to the *Tribune-Democrat* offices, the newspaper's printing press was not functional because of the amount of power it needed to operate. "The press takes an enormous amount of electricity," said publisher Dick Mayer.

> *We had no way to get* [enough] *electric power to the press. So, Jack Murtha sent somebody over, and he said, "Is there anything we can do for you?" We said, "We need a generator," and he got the army reserve to send us a huge generator on tires, one that the military uses. We ran our press from that generator for probably a month until we could get the service that we needed from the electric company.*[226]

On Wednesday morning, Mayer took a circuitous route to get to the *Tribune-Democrat*'s offices. "I couldn't get down the regular way, so I went to Jerome, went to 219, came around the airport and came down Frankstown Road. I drove over to the *Tribune* in about two feet of water," he said. The newspaper publisher had learned of this alternate route from his father, who was Johnstown's district attorney at the time of the 1936 St. Patrick's Day flood. "He had to get to town," said Mayer. "I remember that he told me the only way to get to town [in a flood] was from the Frankstown Hill and the airport. I was able to do the same thing in '77."[227]

In the days following the flood, the *Tribune-Democrat*'s staff found it very difficult to gather and verify information. "We did not know what was going on," said managing editor George Fattman. "The first paper came out, and it said there were just a few people dead. I remember we were in the conference room at WJAC. They had a blackboard, and we would try to write down the names of people we knew were dead or where a body was found. By the end of the day Wednesday, we had next to nothing."

On Wednesday afternoon, Fattman got a call from Bill Dixon, manager of the *Tribune-Democrat*'s bureau office in Ebensburg. "He said that somebody had made it to Ebensburg from Tanneryville," said Fattman. "Somebody had got up there and told him this incredible story that a dam had broken and that there were all kinds of death and destruction in Tanneryville. He said he didn't know if it was true or not but that we should try to get over there, which we did. We really did not know what was going on."[228]

Misinformation also circulated in the city, sending many into a panic. Susan Brett of Cambria City fled for higher ground after hearing a rumor of a dam break days after the rain stopped. "Someone said that the Hinckston Run Dam broke, and we ran to go up the ramp to Brownstown," said Brett, a student at the University of Pittsburgh's Johnstown campus at the time of the flood. "I really thought that I was going to die. We had just seen all of these people die from a dam bursting."[229]

Mayer credited Murtha with helping people in the region get back on their feet. "He got the federal government to do whatever needed to be done," said the *Tribune-Democrat* publisher. "I was on a flood recovery committee for the city and we tried to find out what the problems were and tried to solve those problems. And we called on Murtha to try and solve as many of them as we could." The congressman was also praised by others for his leadership during the crisis. "Jack Murtha was really outstanding," said Gerald Swatsworth, president of Johnstown's U.S. National Bank. "He rolled up his sleeves, he was everywhere, and he got the national guard and other folks in here to give a helping hand. He was an outstanding advocate for the city."[230]

Cambria County judge Caram Abood felt that Murtha, as well as other government officials, responded in a manner befitting the seriousness of the situation. "There were a lot of people who literally lost everything, and they had to have places to sleep, and be fed, and cared for," said Abood. "They started using high school auditoriums and gymnasiums until they could bring in trailers for families to live in temporarily. They had to be provided with food and all the necessities of life. And the government pitched in and helped a great deal, both at the state and federal level. Jack Murtha and the state representatives worked very hard to help."[231]

Reverend Stephen Slavik, pastor of St. Rochus Catholic Church in Morrellville, was also widely praised for his efforts in the aftermath of the flood. A Barnesboro native, Slavik had served as an instructor at Bishop Carroll High School in Ebensburg before assuming his position at St. Rochus in 1966. After the flood, the priest converted his church into a community center, providing food, shelter and showering facilities for many of those who had been displaced. "Father Stephen immediately—that day—had people coming to feed us," said Susan Brett. "He set up a store. He got people to build showers. He was so instrumental with our recovery effort. I don't know what we would have done without him and the parishioners from St. Rochus." Slavik went on to serve as chairman of the city's flood relief center. He hounded lawmakers, including Murtha, to provide help to the people of the city. "He realized that the only way to get their attention, to get them to look at issues fairly, was to confront them," said Larry Olek, one of Slavik's parishioners.[232]

During a break from cleanup efforts in the days following the flood, volunteers enjoyed a party-like atmosphere outside of St. Rochus. The party was organized by Slavik, who provided seventy cases of beer and hired a band to provide entertainment. Volunteers prepared various food items, and there was dancing on streets that had just recently been cleared of mud and debris.[233]

MAINTAINING LAW AND ORDER

Incidents of looting following the 1977 flood were isolated, but they did occur. Thieves raided stores before daybreak on July 20; among the places looted were the Camera Shop on Park Place, James Jewelers on Franklin Street and United Jewelers, Glosser Brothers and the Jupiter Store on Main Street. The Army-Navy Store, the Family Store, the Weiser Music Store and Acme Photo were also looted. "I watched as eight young people ransacked the windows of United Jewelers," said resident Toby Sweeney. "They were stocking up their confederates with items as if on a Christmas shopping spree." At James Jewelers, a group of around ten people smashed a plate-glass window at the front of the store, entered through the broken glass and carried out all that they could.[234]

The looting was not isolated to businesses. "It [looting] went on all night," said one Solomon Homes resident. "Some people are still waiting by their homes with guns." Marine Harvey Freville said many tenants refused to leave their units at the Solomon Homes because they feared their possessions would be stolen. "They barricaded themselves in because they thought if they left, their homes would be looted. The back side of the house might be gone, but they still barricaded themselves behind the door." On Wednesday morning, fifty national guardsmen and police officers were sent to the housing development to protect residents' property.[235]

In parts of the city, citizens were deputized and tasked with policing their own communities. Johnstown mayor Herb Pfuhl issued a "shoot-to-

A number of downtown businesses were looted following the 1977 flood, prompting Mayor Herb Pfuhl to issue a "shoot-to-kill" order to city police officers if they spotted any looters taking advantage of the tragedy. *Johnstown Area Heritage Association.*

kill" order to police officers, and national guardsmen assisted local law enforcement officials in patrolling the streets. "Looters will be shot on sight," Pfuhl said. "We aren't going to have a New York situation here." Anyone arrested and charged with looting was held on $50,000 bail. "We haven't had any trouble with looters here," said an armed security guard on watch at Bethlehem Steel's headquarters on Washington Street, "and we're going to make damn sure we don't have any." The mayor instituted a 10:00 p.m. citywide curfew that remained in effect for a month. He also banned nonresidents from visiting Johnstown.[236]

Pfuhl's orders to the city's police force to "shoot to kill" did not result in anyone being gunned down. The mayor's approach to deterring looters had both supporters and detractors. FBI director Clarence Kelly supported the order, noting that "if it [looting] is permitted to go uncontrolled, it could escalate into attacks against individuals."[237]

The opinion page editors of the *Philadelphia Daily News*, however, disagreed:

> *The floodwaters in Johnstown have brought out more than the lifeboats. They have also flushed out Mayor Herb Pfuhl and his insane approach*

to law and order. Mr. Pfuhl has managed to add to the crisis of the flood by ordering the Johnstown police to shoot to kill looters. Even orthodox Muslims only cut off a thief's hand. Apparently, Two-Gun Herb was too busy cleaning his gun barrels to have absorbed the sorrowful lessons of Kent State and Attica. And he paid no attention to how the New York police handled the looters during the recent blackout. Several thousand were arrested, but not one was killed. The admirable restraint of the New York Police Department was grounded in a simple principle—a principle not applied at Kent State and Attica—life is more important than property.[238]

Some residents accused the Johnstown Police Force of indifference, arguing that the city's law enforcement officials failed to grasp the seriousness of the situation when the flooding began. "Police were cruising around in their cars as the water got higher—they didn't even open their doors," said Louise New, who lived on Washington Street. New said she walked through the rising water and approached a police car parked in her neighborhood. She asked the officer behind the wheel of the car where she and her two children were supposed to go. She said the officer just looked at her and drove off without responding to her question. She and her two children then swam through the cold floodwater from their home on Washington Street to the inclined plane, where they were lifted to safety in Westmont. "I want people to be able to say what really went on down here," said New.[239]

LACK OF COMMUNICATION

Raymond Nasko and his wife were watching the floodwater rise and listening to a police scanner on Tuesday evening. Nasko said it was not until 3:00 a.m. that police discussed any evacuation plans over the scanner. Ray Birch said police cars and boats repeatedly passed by residents in his neighborhood, including many families with children. "The police, mayor, nobody would help," said Birch. "The mayor came down here after 2:00 a.m. to watch the river rise. He was talking to the police about something, but we never knew how bad it was going to be."[240]

Pfuhl issued an emergency warning over the radio and TV airwaves around midnight, but for those without electricity, it did no good. "I couldn't believe what I saw that night," said Birch. "That's our great city, our taxes at work. We sat around here a day and a half before they even told us where to get water." The Department of Environmental Resources estimated that around 30 to 40 percent of residents in flood-damaged regions were without safe drinking water. By Friday, Pennsylvania's Department of Health reported outbreaks of food poisoning among residents who had consumed contaminated water.[241]

One of the biggest problem spots was a broken water supply line between Mineral Point and the Conemaugh and Franklin sections of the city. Murtha contacted Bethlehem Steel and U.S. Steel officials and secured pipe for a temporary supply line. Another line was laid between Westmont and Morrellville to get water to the lower sections of the city. "We have no idea of the time it will take [to restore water service]," said Johnstown Water Authority president Charles Kunkle. "It's our number-one priority." Drinking water was trucked into the valley by the national guard. Several local soft drink companies and other volunteer agencies also provided drinking water until the supply line was fixed.[242]

The national guard was instrumental in the initial cleanup efforts, but military leaders emphasized that their role was a temporary one. "Our people are anxious to return to their homes and civilian jobs," said Brigadier General Robert Carroll, director of the guard's flood relief operations. "Many of the guardsmen from two of our battalions are themselves flood victims and have much to do to attend to their own properties."[243]

Johnstown received $75 million in federal funding for rebuilding and another $25 million was allocated to improve the city's flood control infrastructure. Most of this infrastructure money went to improving drainage and sewage systems along streams in the region. Federal money was also used to improve communication channels between the National Weather Service and Cambria County emergency management officials. Lack of communication was a factor in the loss of life and property in the 1977 flood. The first flash flood warnings were not issued until 2:40 a.m. on Wednesday, July 20. By this time, many sections of the city were already under more than ten feet of water.[244]

The Federal Aviation Agency's weather reporting station at the Johnstown Airport closed at 11:30 p.m. on Tuesday, July 19, and there were no ground reports provided to the NWS throughout the night. This left the agency in the dark on a night when lightning illuminated the valley. The NWS launched an in-house investigation of its operations in an effort to determine whether or not negligence on the part of the bureau contributed to the failure to provide Johnstowners sufficient warning.[245]

Following this internal investigation, NWS officials assigned blame to others outside of the organization. They contended that Johnstown's government leaders were at least partially responsible for the loss of life and property in the disaster. The NWS argued that a flood alert plan, or an automatic alarm system tied to a rain gauge, would have warned residents of the seriousness of the situation sooner. At the same time, however, the NWS concluded that the storm was a "freak occurrence and not forecastable." Meteorologist Dick Mancini had followed the storm on his radar map from the moment it began raining in Johnstown. Typically, weather systems that create thunderstorms move somewhere between twenty and thirty miles per hour before dissipating. However, because of a high-pressure system and no winds in the upper level of the atmosphere above Johnstown on July 19, 1977, a lower storm planted itself over the city and was "refueled" by passing storms. "One thunderstorm would develop in the northwest part of the radar area, move southeast and dissipate; then another thunderstorm would develop," said Mancini. "It was like a conveyor belt."[246]

The flood protection walls that were built after the 1936 flood have, by and large, channeled dangerous rising waters away from downtown Johnstown. On July 20, 1977, however, they were not enough to protect the city and its residents from a third major deluge. *Johnstown Area Heritage Association.*

On this fateful night, Johnstowners discovered that the "controls" put in place in response to the 1936 St. Patrick's Day flood were not sufficient to handle the amount of runoff that the city experienced forty-one years later. Widening the river channels and strengthening their banks with concrete walls to lessen the risk of erosion was enough to prevent the city from suffering any major damage from Hurricane Agnes in 1972. Mother Nature's "blob of precipitation" in July 1977, however, proved to be too much for the reconfigured rivers and their banks, as well as several dams—most notably the Laurel Run Dam—to handle. Roosevelt's declaration that Johnstown, "from now onwards, will be free from the menace of floods" proved premature. A wave of lawsuits were filed against the Greater Johnstown Water Authority, which owned the Laurel Run Dam at the time of the flood, and the Pennsylvania Department of Environmental Resources. After more than a decade of legal maneuvering, all of these cases were settled out of court.[247]

"The only question people had was 'how fast can we clean up?'" said author David McCullough in describing the prevailing attitude of the city's residents following the 1977 flood. "It's a marvelous tribute to the spirit of the people of Johnstown. It happens again and again, but they don't give up."[248]

"We learned that the people who have their roots in Johnstown, the people who have lived all their life here, have a resilience," said Bill Hindman, who lived in the West End since 1945. "They weren't going to get knocked out of the ring. We all pulled together." This resiliency was documented in an editorial in the *Tribune-Democrat*:

> *The history of those earlier major floods—and minor ones—is not composed simply of death and destruction; the history is not complete without the recounting of the massive and successful recovery efforts of those bygone days. So, it was that while the Wednesday waters were receding slightly during the hours heading toward noon, people were beginning to wonder about the difficulties that lay ahead.*
>
> *But the wondering was not cluttered by doubts and despair; the city and its neighboring communities had been through it before, even if few of the people looking on were here even for the 1936 flood. The people, even the young, knew that there was no need for doubt and despair. The cleaning up, the repairing, the getting Johnstown together again, would be accomplished.*
>
> *So, although the Johnstown district has been proved, by nature, not to be flood free, as all of us had believed and hoped, there was the belief—the certainty—that the community and its spirit had been lain awash again but not destroyed.*[249]

6

BETHLEHEM BAILS

F ollowing the 1977 flood, Bethlehem Steel general manager Thomas
Crowley told reporters that the company's Johnstown holdings had
suffered extensive damage. From the car shops in Franklin to the wheel
plant upriver, floodwater had damaged buildings and submerged equipment.
Bethlehem's offices on Locust and Walnut Streets were also damaged.
Bethlehem engineer Jack Hess warned that the damage might exceed what
the company had suffered in the 1936 St. Patrick's Day flood. On July
20, Crowley told a reporter that the roughly eleven thousand Bethlehem
employees around the region should not return to work until notified by
company officials.[250]

On August 1, Bethlehem's bosses gathered for an emergency meeting at
the Sheraton-Carlton Hotel in Washington, D.C. In addition to company
leaders, state and federal officials also attended the meeting, including
Pennsylvania governor Milton Shapp, Senators John Heinz and Richard
Schweiker, Congressman Jack Murtha and Presidential Special Assistant
Greg Schneiders. Bethlehem CEO Lewis Foy had arranged the meeting
after refusing to meet one on one with Shapp the previous week. Foy told the
group that he estimated it would cost his company $35 million to clean up
and rebuild flood-damaged equipment and facilities. The company's CEO
also told attendees that he would demand concessions from union officials
before Johnstown's facilities reopened.[251]

Publicly, however, Foy and other Bethlehem officials were tight-lipped
following the flood, keeping the company's intentions under wraps. "We

don't have any information on that [reopening Johnstown's facilities]," said Marshall Post, Bethlehem's director of public relations. "We're still in the process of cleaning up. We have to evaluate that after an assessment of the damage. Once we determine the extent of the damage, we'll go from there." Johnstown Water Authority president Charles Kunkle flew to the Lehigh Valley and met with Foy several weeks after the meeting in Washington. After talking to Kunkle, Foy approved company funds for cleanup and partial rebuilding efforts of the Bethlehem facilities in the city. However, the Bethlehem CEO told Kunkle that plans to resume work on a basic oxygen furnace in Johnstown were canceled.[252]

At the time of the 1977 flood, the basic oxygen furnace that Bethlehem was building in the city was roughly half completed. Given the unforeseen nature of the flood, Foy was optimistic that state and federal environmental protection officials would allow Bethlehem to continue producing steel through the open-hearth process in Johnstown—if only temporarily. He was right. The Pennsylvania DER granted Bethlehem a five-year waiver on meeting compliance standards in Johnstown, and open-hearth steel production continued in the valley until 1982. In October 1981, Bethlehem erected two electric-arc steel furnaces in Johnstown at a cost of

Bethlehem Steel lagged behind many of its competitors, both in the United States and overseas, in embracing innovation in the steel industry. In its waning years, the company often sat on its hands in regards to the modernization of its operations. In many instances, by the time company officials did act, it was too late to catch up. *Library of Congress, Prints and Photographs Division, Washington, D.C.*

$100 million. The two 185-ton electric furnaces made steel from scrap iron and did not require molten iron. By late September 1982, however, these furnaces were shut down.[253]

The year 1977 is one that lives in infamy in Bethlehem Steel's history. By the end of this year, the company recorded losses totaling $448 million. In January, a three-day blizzard pounded Buffalo, New York, home to the company's Lackawanna plant. The storm shut down this facility and cost Bethlehem $10 million in lost production and repairs. A week later, an underground fire ignited at a Bethlehem coal mine forty miles outside of Pittsburgh. It burned across two of the company's mines for several weeks. This fire caused around $15 million in damages, closed both mines for an extended period of time and forced Bethlehem to buy coal from other companies to power its Pittsburgh-area plants. And then, on July 19, heavy rains began to fall on Johnstown. Given Bethlehem's perilous financial position in the middle of 1977, there was no way that the company was going to invest significant capital in its Johnstown operations, either for full repairs following the flood or for the completion of the basic oxygen furnace.[254]

"The Johnstown flood, while tragic, was the best thing that happened [financially] to Bethlehem Steel that year because it caused us to rethink the upgrades we were planning there," said the company's vice-president Robert Wilkins. "Putting more money into Johnstown would have been like dumping money into a hole, and most of us knew it. That flood saved the company hundreds of millions of dollars."[255]

By 1982, Bethlehem and the rest of the U.S. steel industry had sunk into a deep recession. Since the boom years earlier in the century, more than two hundred plants had closed, and 200,000 jobs had been eliminated. In the span of just five years—from 1977 to 1982—the American steel industry lost $3 billion. The day after Christmas in 1982, Bethlehem announced that it was eliminating 10,000 jobs nationally, including 2,300 in Johnstown. The company consolidated its remaining workers in the city into a unit called the Bar, Rod and Wire Division. Bethlehem officials also demanded that concessions be made by workers who remained employed.[256]

Donald Trautlein, who succeeded Foy as Bethlehem's CEO in 1980, organized labor-management participation teams (LMPT) in Johnstown. Trautlein had joined the company as its comptroller in 1977. Prior to that, he had served as a partner at the Price Waterhouse Company accounting firm, where he oversaw Bethlehem's financial affairs for more than a decade. Initially, his formation of these LMPT teams seemed promising to union officials.[257]

Above: Merry Christmas: roughly 2,300 Bethlehem employees in Johnstown found out on December 26, 1982, that they would be losing their jobs as a result of company downsizing. *Johnstown Area Heritage Association.*

Right: "We regret that conditions have made it necessary to take these further actions," said Bethlehem Steel CEO Donald Trautlein in January 1983, after the company announced it was reducing the salaries of fourteen thousand of its employees. *Courtesy of the National Museum of Industrial History. All rights reserved.*

Trautlein invited Johnstown's mill workers to share ideas on how to improve efficiency. He made the rounds to Johnstown's union halls in the summer of 1983, telling members that if Bethlehem's operations were to continue in the valley, it would need to be under a contract unique to the city. To the surprise of many, union leaders convinced enough of their members to agree to significant concessions.

"Someone did a study that showed jobs that could be eliminated, and the company said that had to happen for the plant to stay alive," said Ron Davies, financial secretary of USW Franklin Local 2635.

> *What they did was combine jobs. For example, where we had a crew with a loader and three chainmen, they came back and said, "We don't need the third chainman. The loader can come by and pick up the slack." So, they eliminate one man on each crew there. Then they go to the grinders complex, where there are four cranemen on duty. They say, "We don't need four cranes running at one time." First, they cut it down to three cranes, and then they cut it down to two. At the Number Three Yard, they eliminated about four chainmen and picked up the slack by making the inspectors and loaders and whoever else was in the yard take turns helping chain. They just doubled up jobs.*[258]

Davies's union hall rejected Bethlehem's proposed plan on the first vote. Company officials told union leaders it would terminate its operations in Johnstown altogether if its proposal was voted down a second time. Facing this threat, all six union halls in the region voted to accept the concessions. When terms of the agreement were announced, dissension swept throughout Johnstown's Bethlehem workforce. Many argued that the "yes" votes had been bought by offering lucrative pensions and lowering retirement-age and years-of-service qualifications. "The older workers voted for it just to get out, and a lot of young people on layoff voted for it just to get back to work," said Davies. Four of Johnstown's union presidents were voted out of office in the months after the agreement.[259]

Trautlein demanded more cutbacks in the valley a year and a half later. The company had suffered a fourth straight year of significant financial losses in 1984, with its Johnstown division struggling mightily. The Bethlehem CEO told union officials that no more money would be invested in the city unless it demonstrated a profit by the end of the calendar year. It did not.[260]

On March 23, 1985, the six locals in Johnstown's Bar, Rod and Wire Division reached a cost-savings agreement described by former *Bethlehem*

Globe-Times editor John Strohmeyer as "probably the most revolutionary [agreement] ever reached between a plant and a major steel company." Terms of the agreement included:

1. Pay cuts: Hourly rates were reduced by eighty-two cents per hour and a wage increase of forty-five cents per hour that was set to begin in February 1986 was canceled. In place of these lost wages, mill workers were given preference stock through an employee investment program that was equal to the lost revenue.
2. Elimination of incentive pay: Extra wage incentives, based on output, were discontinued. In their place, a profit-sharing plan was instituted.
3. Fewer vacation days: Bethlehem employees who had accrued more than three weeks of vacation eligibility lost two weeks. One week was lost for employees with two weeks of vacation time to their credit.
4. Fewer holidays: Ten annual holidays were reduced to seven. In their place, preference stock equal to the lost vacation days was provided to employees.
5. Less healthcare coverage: Visual and dental coverage were canceled.
6. Lower shift premiums: Increased wages of thirty and forty-five cents per hour—for those who worked second and third shifts, respectively—were reduced to ten and twenty-five cents. Holiday premiums were decreased, and Sunday premiums were eliminated. Bethlehem leaders forced members of management to accept concessions as well, reducing salaries by $845,000 annually across the board. The average of these pay cuts for salaried employees was $1,706 per year.[261]

"A couple of years ago, the biggest fear in this town was the rain," said Jack Sabo, treasurer of Franklin Local 2635. "Every time it rained, we worried whether the South Fork Dam would break. Now, the big worry is when someone says, 'I hear Trautlein is coming to town.' It's the same damn thing—the dam is going to break, or Trautlein is going to make more cuts. The town is running scared one way or the other."[262]

In August 1992, Bethlehem Steel sold its last remaining plant in Johnstown—the Gautier Rolling Mill—to J-Pitt Steel, a Pittsburgh company owned by Michael Pitterich. The city's wire mill was also purchased and renamed Johnstown Wire Technologies. Three months later, it appeared that Bethlehem had reached an agreement to sell Johnstown's Bar, Rod and Wire Division—as well as its holdings in Buffalo, New York, and Sparrows Point, Maryland—to the Ispat Group based in Calcutta, India. The deal with Ispat ultimately collapsed, however, because of opposition from labor

After decades of employing hundreds of thousands of Johnstowners, Bethlehem Steel sold the last plant it was still operating in the city—the Gautier Rolling Mill—in August 1992. For all intents and purposes, the steel era in Johnstown was over. *Johnstown Area Heritage Association.*

leaders. In December 1993, Bethlehem sold the Bar, Rod and Wire Division to Moltrup Steel Products.[263]

In October 2001, Bethlehem Steel filed for bankruptcy protection, citing the 9/11 terrorist attacks as the event that pushed the company over the brink. "Since September 11, what was a declining marketplace has been a freefall," said Bethlehem CEO Steve Miller. He said sales had gone to "hell in a handbasket." Miller—who had helped lead an economic turnaround at the Chrysler Corporation—had assumed the top job at Bethlehem a month earlier.

Bethlehem became the twenty-fifth U.S. steelmaking company to file for bankruptcy protection in just three years (1998–2001). At the time of its bankruptcy announcement, the company had 13,000 employees across its remaining holdings and 130,000 retirees receiving a company-paid pension plan and healthcare benefits. Since the start of 1998, Bethlehem had sustained a staggering $1.6 billion in losses. The company claimed to have $4.2 billion in assets and $4.5 billion in debt when it filed for bankruptcy in October 2001.[264]

In December 2002, the federal government assumed responsibility of Bethlehem's pension payments to ninety-six thousand retirees from the company. The steel company could no longer fulfill them. The government immediately put an end to a longstanding Bethlehem policy that allowed company employees to retire after thirty years of service. "I am very sorry for their [employees'] personal situations. This is not how the company wanted it," said Miller. "We wanted to have more time so as to be able to reduce our workforce by special early retirement programs, so those people who were not going to continue working would have the benefit of an early retirement. That was not in the cards. The government slammed the door in our face. We're upset about it, but it is the reality. We do not have the money to make good on all of the promises made by this corporation over the last fifty years."[265]

On February 8, 2003, the Bethlehem Steel Corporation, which had once ranked eighth on the Fortune 500 List, passed into history. Its final half-dozen plants were sold to the International Steel Group for $1.5 billion.[266]

EPILOGUE

We kind of got spoiled with Bethlehem Steel. When I went to high school, my father worked in the steel mill. He worked there for forty-six years. My whole neighborhood took vocational courses, and we were going to go into the mills. Because that's what you did. If you didn't have money to go to college, you went into the mills. The family I grew up in didn't have money to send me to school, so I was going to work in the mills. What happened was, I had a business guy who got interested in me, talked to me about school and helped me get a scholarship. So, unlike my buddies, I ended up going to college. But, you know, there was nothing wrong with that [going into the mills]. *You get in the mills, you had a decent paying job, you had benefits, you had retirement, you could raise a family. The cost of living was not bad. With Bethlehem leaving—much like lots of other towns—Johnstown probably never recovered from that.*

—Donato Zucco, mayor of Johnstown (1998–2006)[267]

MOVING FORWARD

The story of Johnstown, Pennsylvania, is similar, in many ways, to those of other Rustbelt communities across the Northeast and Mid-Atlantic regions of the country. For many generations, Johnstowners rode the waves of the economic boom and bust periods in the steel and coal industries. In Johnstown, however, its citizens also dealt with the constant threat of

After three major floods and the steel industry's demise in the city almost forty years ago, Johnstown is still trying to regain its economic balance. *Library of Congress, Prints and Photographs Division, Washington, D.C.*

floodwaters. Following the third major flood to hit the city in 1977, the water eventually dried up—but so did the jobs. Another boom for the steel industry was not on the horizon. From the middle of the twentieth century to the turn of the twenty-first century, Johnstown lost four-fifths of its manufacturing jobs. Only Steubenville, Ohio, suffered as traumatic a loss to its manufacturing economy over this same period.[268]

Johnstown's economic challenges continue today, almost a half century after the 1977 flood and the silencing of Bethlehem Steel's open-hearth furnaces in the city. Johnstown has been mired in a seemingly endless retrenchment period, unable to identify what its second act is going to be. Steel had served as the city's economic engine since its infancy, and Johnstown has struggled since this lifeline was severed. At the end of 2019—two months before the onset of the global pandemic—Johnstown's job growth rate was -1.1 percent. The cost of living in the city was 16 percent below the national average, and the median household income was around $49,000.[269]

Young people continue to leave the city in large numbers, and the consequences of the opioid epidemic in the region are still not fully understood. According to a 2019 study conducted by *24/7 Wall Street*, Johnstown ranked second in the nation on a list of the fastest shrinking cities from 2010 to 2018, losing 8.2 percent of its population. Johnstown also saw a -12.3 percent unemployment change during this period. This decline can be attributed, in large part, to the deaths of older residents—many of them former mill workers and coal miners—as well as the exodus of young people from the region. The city has seen its population plummet from forty thousand in 1977 to around twenty thousand in 2021.[270]

Randy Frye serves as dean of the Shields School of Business at Saint Francis University in Loretto, Pennsylvania, the hometown of former Bethlehem Steel president Charles Schwab. Frye has lived in the Johnstown region since the mid-1950s. His office on Saint Francis's campus is on the first floor of Schwab Hall, which is named after the former steel leader. Frye was on break from college when a third major flood decimated his hometown in July 1977. At the time of the flood, he was managing his father's gas station in Johnstown. The station was forced to close for five months due to flood damage to the road in front of the business, so Frye accepted a job with Johnstown's community public works program. He spent the remaining weeks of his 1977 summer break from college cleaning mud out of people's homes and businesses in the Moxham and Hornerstown sections of the city.

Frye is an active member of the Johnstown Area Regional Industries (JARI) group. He believes that it is going to take a diversified approach to revive Johnstown's economy, an approach that extends beyond the city:

> *We need a Cambria County strategy, not just a Johnstown strategy. It is going to take a lot of small businesses opening and being successful. We need family-sustaining jobs. We have a good healthcare infrastructure, but it is not going to be one thing. It needs to be a lot of different pieces coming together—education, healthcare, technology and remote work, combined with what's left of defense. All of our eggs were in one basket. Bethlehem Steel was King Kong. When King Kong left, we didn't have a replacement.*[271]

The primary non-healthcare industry in Johnstown today is high-tech defense, a legacy of the Jack Murtha era. The congressman and legendary legislative dealmaker brought much-needed federal money to his district in the aftermath of the 1977 flood in the form of cleanup assistance and state and federal grants and loans. He continued to deliver federal grants and jobs to the region in the years that followed. Today, more than a decade after Murtha's death in 2010, several businesses in the valley continue to fulfill sizable government contracts, and Johnstown remains home to the annual "Showcase for Commerce," one of the nation's premier defense trade shows.[272]

Nothing, however, has come close to providing the economic stability that steel afforded generations of Johnstowners. This is a consequence of the city being tied to a single industry since its founding. No major industry aside from healthcare has been able to sustain itself since Bethlehem Steel shut down its operations. And many of the federal contracts that Murtha helped steer to his legislative district have dried up since his death. It is estimated that Murtha helped deliver as much as $2 billion in federal money to Johnstown between 1995 and 2010.

In an article written by *New Republic* columnist Jason Zengerle at the time of Murtha's death, the author suggested that the city was as ill-prepared for the legislator's passing as it was for Bethlehem's demise:

> *Murtha's death signals something more than the death of a man or the death of an era: It likely spells the death of the city he represented. When Murtha was alive, Johnstown raised myriad monuments to him—placing his name on everything from a technology park to an airport. But the city never prepared itself for the day when its honors to Murtha would have to come in the form of memorials. Johnstown's success was not a façade, but its prosperity was as dependent on one congressman as it had once been on one industry. It was almost as if Johnstown could not bring itself to imagine—and thus prepare for—what would happen once Murtha, like steel before him, was no longer there to sustain it. And now, it will face the consequences of that failure.*[273]

Frye has a less jaded view of Murtha's efforts than Zengerle. He believes that the former congressman's intentions were good but that they were thwarted in many instances by poor management:

> *Murtha's vision was not just to give pork. His vision was to spread around seed capital, with the idea that these companies would become sustainable industries. In many cases, they did not because of ineffective management. I don't think he saw it as a temporary fix. He was trying to give us a "reboot" around defense. DRS [Leonardo DRS] and CTC [Concurrent Technologies Corporation] and Lockheed [Lockheed Martin AeroParts] are well-managed, they have been sustainable and they still*

get contracts. They took root. But many others did not, and I don't know if
we can blame Murtha for that. I don't think the beneficiaries of Murtha's
connections served him well. They didn't have imagination and vision. A
few of them were able to break away from the apron strings but not enough
of them. A lot of people behaved in a way where they thought that they
could just continue living off the government dole.[274]

Despite three major floods, Bethlehem Steel's demise, Murtha's death, an opioid epidemic and a global pandemic, the city's spirit of self-preservation has not waned. Just as businessmen Louis Galliker, Daniel Glosser, Frank Pasquerilla, Walter Krebs, Robert Gleason, Andrew Koban, Charles Kunkle, Howard Picking and others collaborated to form JARI in June 1974 in an effort to preserve steel production and identify other avenues for economic growth, Johnstown's civic leaders of today are working together in a collective effort to steer the city toward a brighter economic future.

JARI has collaborated with numerous groups, including Vision Together 2025, the Community Foundation for the Alleghenies, the Johnstown Redevelopment Authority, the Cambria County Chamber of Commerce, and the Discover Downtown Johnstown Partnership, among others, in an effort to reverse the city's economic trajectory.[275]

There have been some successes. The downtown has been revitalized over the last decade, and many of the small businesses operating there were enjoying a rebirth before the global pandemic. JARI has also tried to capitalize on the city's history through tourism, promoting outdoor opportunities in this beautiful section of the Laurel Highlands Mountains. Currently in development is a two-mile walkway that will be named the "Iron to Arts Corridor." It will follow the path of the Conemaugh River. Tangible progress has also been made in the city in the areas of public health, education and blight removal.[276]

It remains, however, a steep climb out of a deep hole for Johnstown. By most measures, business development has languished in the city over the last half century. When Bethlehem's open-hearth furnaces stopped firing, the city lost its course. It has not been able to navigate a new path to sustained economic viability. Johnstown, of course, is not alone. There are many former Rustbelt cities across the nation struggling to reinvent their economies in the wake of steel's decline a half century ago. These cities' struggles are well documented. But there are also some success stories of former mill towns that have reinvented themselves, and these stories could, perhaps, provide a blueprint for Johnstown's government and business leaders.

Bethlehem, Pennsylvania—the former base of operations for Bethlehem Steel—has transformed itself and its economy in the years since the company's furnaces stopped firing. *Tim Kiser, Wikimedia Commons.*

One of these success stories is Bethlehem, Pennsylvania, the former headquarters of Bethlehem Steel. The city stretches over nineteen miles in eastern Pennsylvania's Lehigh and Northampton Counties. In 2020, Bethlehem's population of 76,370 ranked seventh in the state, and the city enjoyed an annual growth rate of more than 0.5 percent in each of the three years before the pandemic (2017–19). Bethlehem's population is larger today (75,961 in 2021) than it was in the middle of the last century (66,340 in 1950).[277]

Dale Falcinelli is a longtime faculty member in the College of Business at Lehigh University, which is located in Bethlehem. He has provided consulting services to numerous businesses and organizations in the city and throughout the Lehigh Valley for more than four decades. Falcinelli believes that most of the credit for Bethlehem's economic transformation after the mills closed should go to the city's business and government leaders:

> *Bethlehem's rebirth was very slow and very deliberate. You had great strategic planning and leaders in the community who understood what the impact was. If you pull the plug on the area's largest employer, you are going to stop dead in your tracks. Across time, the leadership decided to take a hard look at the community, and reinvestment began to happen in the service economy and other areas. It was like time-lapse photography—eventually, we began to see the flowers bloom.*
>
> *All of this wonderful leadership—that's what turned the corner. We had the right leaders. And let's not forget the Lehigh Valley's wonderful location, one that eventually became a drawing card for distribution-based companies. The Lehigh Valley Economic Development Corporation became a driving force, one responsible for the economic reprogramming of the valley, in an alliance with leadership in the community. It was amazing.*

*You look back and you think, "They fell out of bed, and it just happened."
No, what created this vibrant community of ours are people who had belief
in themselves and this community. And all of a sudden, the dots started to
connect. Lehigh had some strategic advantages that were leveraged. Today,
it's a service and high-technology economy.*[278]

Foremost among Bethlehem's strategic advantages is its location and
highway system. The city has easy access to major highways in every
direction and is less than an hour-and-a-half drive from New York City
(to the north), Philadelphia (to the south) and Harrisburg (to the west); 40
percent of the nation's population is within an eight-hour drive of the city.
Johnstown, conversely, has arguably the worst highway system of any city
in the state.

A significant allocation of property tax revenue toward redevelopment has
also helped power Bethlehem's recovery following the collapse of the steel
industry. On the 124-acre site where Bethlehem Steel's plants once operated
is the Wind Creek Bethlehem Casino, which generated $521 million in profits
in 2018 and $522 million in 2019. Wind Creek raked in the second-highest
profits of any of Pennsylvania's twelve casinos in 2019. The Keystone State
ranked number one in the nation in tax revenue from the casino industry in
2019 ($1.5 billion). The Ben Franklin Technology Partners, an initiative of
the Pennsylvania Department of Community and Economic Development,
has also played a key role in transforming Bethlehem from a manufacturing-
based economy to a technology-based one.[279]

Falcinelli believes that Bethlehem's transformation could serve as a model
for other former steel towns, like Johnstown, that are still struggling to find
their economic footing in the wake of the industry's decline:

*When something this severe happens, it tests the community's leadership to
engage in strategic planning. And I am not throwing stones at Johnstown,
but if you just go after federal money and it is thrown at you by the bucket-
full, that's no solution. It is an inoculation that is short-lived in terms of
the benefit. Federal money can be very poisonous. Too much of it becomes a
distraction for a community because it appears to be a quick fix.*

*Adverse times test leadership. Would we have a community the likes of
which we have today if we didn't have to engage in the struggle, given the
death of Bethlehem Steel? The good to come out of the death of Bethlehem
Steel was a community that reached down, girded its loins and did what
needed to be done. There was a vision, there was a strategy and there was*

a commitment to getting results over time. When Bethlehem Steel pulled the plug, it would have been easy to walk around the community and say, "It's over, it's done, it's finished." We rose from the ashes, but it was driven by survival.

This wasn't about incremental growth, about getting back on track—it was about transformation. The Lehigh Valley has been transformed. Transformation requires a different type of leadership: a belief in oneself and a belief in what can be, and that's what we had. The death of Bethlehem gave new life and opportunity to the Lehigh Valley.[280]

LEARNING FROM THE PAST

The third major flood that hit Johnstown in 1977 was not the "event" that triggered the city's downward economic spiral. Bethlehem Steel's plants in the city and in its other locations had been declining for years. The company's decline can be traced to a number of factors, including significant increases in imported steel and the company's own mismanagement. Bethlehem's mills were aging by the 1970s, few large-scale capital improvements had been pursued, and management was paying employees unsustainable wages and offering them expensive benefit programs that had been secured by the unions. These wage increases and benefit packages exacted a serious toll on Bethlehem's bottom line. The 1977 flood factored into none of this, other than serving to accelerate the decline of the company's operations in Johnstown. According to Randy Frye, flood or no flood, Bethlehem's bust and the region's subsequent economic decline were inevitable:

It [the 1977 flood] gave Bethlehem Steel a good reason to leave, but the tsunami of change hit the steel industry long before 1977. What made Bethlehem Steel go bankrupt wasn't a rival company. It was other forces. It was the viability of substitutes. It was the increase of the availability of substitute products for steel in car manufacturing and construction: plastics, ceramics, aluminum. And there were so many advantages to these substitutes. All of a sudden, the buyers of steel had choices. You also had lower barriers of entry. You had foreign steel coming into the United States. You had the mini mills, which were using scrap steel and making steel at a fraction of the cost.

These large, integrated dinosaur plants that you saw in Birmingham, Sparrows Point and Johnstown—they were doomed to failure. And I'm not antiunion, but the U.S. steelworkers negotiated very good contracts: steelworkers had sabbaticals, weeks of vacation, high wages, defined retirement plans. Bethlehem became a bloated bureaucracy. It became complacent, somewhat arrogant and myopic and took a short-sighted view. They fell in love with their way of doing things. They didn't adapt. And if you don't adapt, you die.[281]

Johnstown and its steel industry were inextricably linked to one another, dating back to the founding of the Cambria Iron Works. Cambria Iron, Cambria Steel, Midvale and Bethlehem delivered the jobs, and Johnstown's workforce delivered the labor in a mutually beneficial relationship, one that lasted for more than one hundred years and one that sustained the city's economy through the Civil War, two world wars and three major floods.

Cyrus Elder lost his wife and daughter in the Great Flood of 1889. He delivered a speech in 1900 on the one hundredth anniversary of the founding of Johnstown. A native of Somerset, Elder was hired by Cambria Iron Works president Daniel Morrell and served as chief counsel at the plant following his military service during the Civil War. Elder later served as associate editor of the *Johnstown Tribune* and was very active in city government. His centennial speech in 1900 proved prophetic:

What are the changes which may be looked for in the next one hundred years? Will railroads be superseded by some new methods of transportation? Can the shares become as valueless as turnpike stock? Will steel be replaced by some lighter and stronger and cheaper metal? Will steam and compressed air and electricity give place to some force which creates itself and costs nothing, like solar energy, the ocean tides or the pressure of the atmosphere? Will there be a different kind of man and a new social and industrial order?[282]

Cyrus Elder's wife and daughter were killed in the 1889 flood. The family's home was destroyed by the floodwater. *Johnstown Area Heritage Association.*

In the century following Elder's speech, the landscape did change drastically for the steel industry in Johnstown and other mill towns around the country. Some of these changes were inevitable, due to increased competition, unsustainable wages and benefits and the higher cost of doing business in a nation that was becoming increasingly concerned with corporate pollution. Yet while all of these factors played a role in domestic steel's precipitous decline over the last quarter of the twentieth century, many of Bethlehem Steel's wounds were self-inflicted, both in Johnstown and across its other holdings.

The company's leaders—over a span of decades—failed to embrace new technologies. They blamed lawmakers for their economic woes, and this complaining was not unique to Bethlehem—officials at almost every American steel company lobbied for higher tariffs to be placed on imported steel. They attributed their financial problems, by and large, to what they perceived as government leaders' unwillingness to protect their interests.

Amid the rising challenge presented by international steel companies, however, Bethlehem and many of its domestic steel-producing brethren failed to make meaningful changes to their own operating procedures, falling further and further behind. Bethlehem's bosses naively believed that because the company had played a crucial role in the growth and successes of the country throughout much of the century, the federal government owed it a degree of help and protection. This miscalculation played a significant role in the steel giant's demise.

What officials at Bethlehem failed—or were unwilling—to acknowledge during the second half of the twentieth century was that international competition was not the driving force in the company's decline. By the 1970s, Americans did not use as much steel as they had when the Cambria Iron Works was powering Johnstown's economy or when Bethlehem Steel was bolstering the Allies' war effort. Substitute materials were increasingly being used in place of steel. In addition, Bethlehem officials failed to commit to research and development, failed to anticipate other market challenges and failed to identify new paths for growth. The company failed on these fronts while paying its workforce unsustainable wages and providing its retirees unmanageable benefit packages. Steel imports and floodwaters in Johnstown factored into none of these many missteps.

Regarding research and development, Bethlehem doubled down on Andrew Carnegie's sentiment that "pioneering don't pay." While the company did build a $10 million research laboratory in 1961 at its headquarters in Bethlehem, most of the work at this "lab" focused on day-to-day problems. In 1945, the company's vice-president James Slater noted that "unlike some [companies], who can't wait, we don't have to add capital unless we are sure it will pay out." In far too many instances over the second half of the twentieth century, by the time Bethlehem officials committed resources to an opportunity, it was too late—competitors in the United States and abroad had beaten them to the punch. "We move only when improvements are so good we can no longer afford what we've got," said another Bethlehem vice-president, John Jacobs, in 1962.[283]

Bethlehem's unwillingness to embrace the basic oxygen furnace is an example of the company's flat-footedness regarding ingenuity. The company had invested $250 million in the open-hearth steelmaking process during the 1950s. A decade later, however, this method was outdated. The early models of basic oxygen furnaces produced two hundred tons of steel from iron in forty-five minutes. Open-hearth furnaces needed four hours to produce the same tonnage. The basic oxygen furnace was invented in 1952, but Bethlehem did not begin to seriously invest in this technology until the middle of the 1960s. In Johnstown, Bethlehem did not begin constructing a BOF until 1974, before the project was scrapped following the 1977 flood. While Bethlehem operated its far-less-efficient open-hearth furnaces through the 1960s and 1970s, German and Japanese steelmakers were churning out high-quality, more affordable steel produced through the BOF process.[284]

Two of the primary reasons Bethlehem failed to move quickly when it came to technological improvements were the company's vertically integrated structure and its insulated environment at the management level. Members of management almost always rose from within the company's ranks. Few had worked for other businesses, let alone other steel companies. This contributed to a groupthink dynamic at Bethlehem, one that stifled ingenuity. The kind of creativity that Charles Schwab had brought to the company decades earlier was conspicuously lacking by the middle of the twentieth century. "You have a group of [Bethlehem] execs—one of the highest-paid groups in the 1960s," said Dale Falcinelli. "This group that inherited Bethlehem and its rich history was a group that thought that life in the future would be pretty much the same as life in the past."[285]

The higher wages and expanded benefit packages that steelworkers secured through the success of the labor movement also proved to be very costly to Bethlehem and its long-term viability. "I think the crux of Bethlehem's problems started back in 1959 with the strike," said Charles

Bethlehem Steel is long gone, but its shadow looms large over the city in the valley. *Johnstown Area Heritage Association.*

Luthar, who worked at Bethlehem for thirty-three years. "That set the tone for the way Bethlehem operated from that point on." Noted Bethlehem Steel historian John Strohmeyer: "Some of the contracts were outrageous. On the other hand, steel management was under tremendous pressure by government to settle—they didn't want any more strikes. They were caught in the middle. The union has got to take some blame, as well as the management."[286]

In an ironic twist, Johnstown's Great Flood of 1889 played a role in the rise of the U.S. labor movement. In response to Johnstowners' lack of success in collecting any damages from members of the South Fork Fishing and Hunting Club following the break of the South Fork Dam, state and federal courts began to weigh more heavily the consequences of the actions (or inaction) of business leaders on the working class. This included an increase in the application of the Rylands Rule in U.S. mill and mine accidents in the decades following the Great Flood.

Rylands v. Fletcher was a legal case in Lancashire, England, during the 1860s. It involved mill owners who had constructed a reservoir on their land. Water in the reservoir broke through the shaft of a neighboring abandoned coal mine and flooded several connecting passageways in the privately owned mine. In 1865, a trial determined that the reservoir owners were free of any negligence because they were not aware of the abandoned mine shaft. The mine owner appealed this decision, however, and a higher court in England overturned the lower court's ruling in 1868, finding favor with the plaintiff. "The true rule of law," noted British Justice Colin Blackburn, "is that the person who, for his own purposes, brings on his lands and collects and keeps there anything likely to do mischief if it escapes, must keep it in at his peril, and if he does not do so, is prima facie answerable for all the damage which is the natural consequence of its escape."[287]

In the decades following the 1889 Johnstown flood, courts across the United States began to impose stricter penalties on steel and coal companies, railroad owners and other business leaders when their companies' activities resulted in damage to property or loss of life among their workforce and in their communities. This legal protection helped set the wheels in motion for the rise of the U.S. labor movement, the eventual success of which would exact a serious toll on Bethlehem Steel's bottom line and Johnstown's economy. In addition to an increased application of the Rylands Rule, the Great Flood was also a catalyst for other new rules designed to protect working-class Americans, including the Sherman Antitrust Act of 1890. Consequences from the Great Flood also factored into President Theodore Roosevelt's decision to break up monopolies during the early twentieth century.

LOOKING TOWARD THE FUTURE

While the return of large-scale steel manufacturing in Johnstown is highly improbable—if not impossible—the same cannot be said of future floodwaters. "It has been flooding in Johnstown for millions of years," wrote David McCullough. "The location of the city that once drove America's steel industry continues to leave it vulnerable to another catastrophic flood."

The flood controls that were put in place following the 1936 flood were not enough to withstand the type of runoff that Johnstown experienced in July 1977. Nor are there any guarantees that adequate protections are in place to guard against future floods. As recently as September 2021, more than two thousand people were evacuated from the South Fork area after water contained in the Wilmore Dam rose to within eighteen inches of the top of this earthen structure following heavy downpours in the wake of Hurricane Ida. Indeed, it is impossible to say with full confidence that the Flood City is flood-free moving forward.

The one thing that we can say with certainty is that there is a survivor's spirit in the city in the valley, one that neither floods nor silenced blast furnaces can extinguish.

NOTES

Introduction

1. Hutcheson, *Floods of Johnstown*, 1–2.
2. "25 Die in Flood: Death List Seen as Growing," *Johnstown Tribune* (hereafter *Tribune*), March 20, 1936; "City's Second Major Deluge Strikes Area," *Tribune*, March 20, 1936; House Committee on Flood Control, Hearings on Levee and Flood Walls, Ohio River Basin; Kozlovac, "Adventures," 14.
3. "Army Engineers Study Dam Sites Suggested by State," *Tribune*, January 20, 1938; Kozlovac, "Adventures," 14; Whittle, *Johnstown, Part 1*, 249–51.
4. Franklin Delano Roosevelt letter to Walter Krebs, March 6, 1944, Johnstown Area Heritage Association, Johnstown Flood Collection, 1881–1977; Sid Weinschenk, *Flood Free Johnstown* (Johnstown, PA: Flood Free Johnstown Committee, 1943); Gary Rothstein, "Four Years After '77 Flood, Johnstown Is Better Prepared," *Tribune*, August 16, 1981.
5. Reutter, *Sparrows Point*, 413; Assad, Frassinelli, Venditta and Whelan, *Forging America*, 128, 148; William Serrin, "Steel Industry Woes Weigh Heavily on Johnstown," *New York Times*, October 3, 1982.
6. Rothstein, "Four Years."

Chapter 1

7. "The Great Calamity," *Harper's Weekly*, June 15, 1889.
8. Harper, "Town Development," 71.

9. Kozlovac, "Adventures," 8; Toner, "Floods," 14; McCullough, *Johnstown Flood*, 15.

10. Shappee, "History of Johnstown"; Wallace, "Character."

11. Dickens, *American Notes*, 182.

12. Shappee, "History of Johnstown."

13. Wallace, "Character."

14. Cambria Iron's former general office on Washington Street is still standing today (Toner, "Floods," 8).

15. McCullough, *Johnstown Flood*, 51.

16. Ibid.

17. Frank, "The Cause," 63–66; McCullough, *Johnstown Flood*, 41.

18. Toner, "Floods," 5.

19. Kozlovac, "Adventures," 8; McCullough, *Johnstown Flood*, 43.

20. Nasaw, *Carnegie*, 24.

21. Oddly, Henry Clay Frick's daughter Martha also died at Braemar on a visit to Carnegie's mansion. (Donnelly, *Buildings*, 540).

22. McCullough, *Johnstown Flood*, 51.

23. Wallace, "Character."

24. Roker, *Ruthless*, 51.

25. Shappee, "History of Johnstown."

26. McCullough, *Johnstown Flood*, 73–74.

27. Ibid., 74–75.

28. Michael Novak, "125 Years Later: The Johnstown Flood," *Huffington Post*, May 28, 2014; Toner, "Floods," 6.

29. Frank, "The Cause," 63–66.

30. National Park Service, "Emma Ehrenfeld Statement," www.nps.gov.

31. Toner, "Floods," 8; Hutcheson, *Floods of Johnstown*, 2.

32. Toner, "Floods," 8.

33. Ibid.

34. "Hundreds of Lives Lost," *New York Times*, June 1, 1889; Hutcheson, *Floods of Johnstown*, 1–2.

35. Anna remarried and moved to Richmond, Virginia. She died in 1928. Her grave is marked with a large monument in Johnstown's Grandview Cemetery.

36. "A Valley of Death," *Three Rivers Tribune*, June 7, 1889.

37. Toner, "Floods," 9.

38. Hutcheson, *Floods of Johnstown*, 1–2.

39. "Valley of Death," *Three Rivers Tribune*.

40. The attention and praise that Hastings received in leading the cleanup efforts following the flood helped thrust him into the statewide political scene. He

lost the Republican nomination for governor to George Delamater in 1890, before running for the office again and landing his party's nomination four years later. In the 1894 gubernatorial general election, Hastings defeated Democrat William Singerly (Toner, "Floods," 11).

41. Barton took up residence at 662 Main Street, the home of John Ludwig, during her time in Johnstown.

42. Peter Smith, "Johnstown Flood of 1889," *Pittsburgh Post-Gazette*, May 24, 2014.

43. Roker, *Ruthless*, 206.

44. "Book of the Johnstown Flood," *Tribune*, March 17, 1936.

45. Frank, "The Cause," 63–66; In his 2019 book, *Johnstown's Flood of 1889*, author Neil Coleman provides evidence that the club caused the dam's break by lowering the height of the structure by three feet. Coleman speculates that the club's members suppressed and modified the report of the investigators, who belonged to the American Society of Civil Engineers (Roker, *Ruthless*, 206–7).

46. Eric Pace, "Frank Shomo, Infant Survivor of Johnstown Flood, Dies at 108," *New York Times*, March 24, 1997; Shomo made his livelihood as a foreman for the Pennsylvania Railroad for five decades. (Roker, *Ruthless*, 263.)

Chapter 2

47. Whittle, *Johnstown, Part 1*, 18.

48. Berger and Glosser, *Johnstown*, 461.

49. Whittle, *Johnstown, Part 1*, 10.

50. Strohmeyer, *Crisis*, 196; My grandfather John Earl Boland "Pappy" worked as a laborer in Cambria County coal mines from 1932 until his retirement in 1970. He served as an underground laborer and cutter for the Monroe Coal Company from 1932 to 1948. On October 15, 1948, the Bethlehem Steel Company bought Monroe Mining. Pappy worked as a cutter and shuttle car operator for Bethlehem from 1948 until 1970. In the fall of 1962, he was working at Bethlehem Mine No. 32 in a shaft beneath the village of Revloc, a community roughly twenty miles north of Johnstown. He was operating a shuttle car, a buggy that transports coal extracted from underground seams, to a main belt, where it is carried out of the mine. Pappy suffered a cervical fracture in his back when a rock fell on him from the ceiling of a section of the mine where he was working. The rock knocked him into his buggy. He was carted to the surface by

fellow miners and taken to Conemaugh Valley Memorial Hospital in Johnstown, where doctors drilled holes into each side of his head. Pappy was placed in cervical traction for six weeks. He had to relearn how to use his hands and fingers. When he returned to his home in Ebensburg, he had to drag around one of his feet, which was temporarily paralyzed. He couldn't lift his arms above his shoulders. Like many miners who had suffered serious injuries on the job, he needed to design and implement his own physical therapy regimen. He practiced falling; relearned how to use his hands by hitting a punching bag; strengthened his upper body with a weighted pulley system; and worked to regain the use of his fingers by squeezing tennis balls. Perhaps his most creative rehabilitation endeavor was an electrical stimulation device he created to help generate nerve growth in his back. The components of this device included a piece of wood that was about two and a half feet long with two cake pans (attached to the board) and electrical wiring. He placed damp towels over the cake pans, plugged the electrical cord into an outlet and touched the bare wires at the other end of the cord to one of the cake pans. The electricity that was generated shocked him repeatedly. Pappy eventually regained the use of his fingers when the nerve endings regenerated. He returned underground and to the Bethlehem mines six months after his accident. ("A Brief History of Immigration and Migration to Johnstown," Johnstown Area Heritage Association, www.jaha.org.)

51. *History of Appalachian Coal Mines*, 11.
52. Ibid.
53. Ibid.
54. Pennsylvania Bureau of Mines Annual Report, 1900.
55. Steven Pavlik, "Disaster in Johnstown," *New York Times*, July 11, 1902, 1.
56. "A Brief History," Johnstown Area Heritage Association.
57. Immigrants in Industries, Part 2, Immigration Commission, 32, 33; James Swank quoted in J.B. Lippincott, *Progressive Pennsylvania* (Philadelphia: J.B. Lippincott Company, 1908), 7; Whittle, *Johnstown, Part 1*, 44.
58. McDevitt, *Banished*, 23.
59. Nelson Raynor, "Johnstown's Colored Population," *Johnstown Weekly Democrat* (hereafter *Weekly Democrat*), October 30, 1903.
60. "Ku Klux Klan Holds Secret Meetings Here," *Tribune*, January 27, 1922; Jenkins, *Hoods and Shirts*, 71–72; McDevitt, *Banished*, 23.
61. Sherman, "Johnstown v. the Negro," 454.
62. "Midvale Shows a Net Loss," *Tribune*, April 3, 1923.
63. Marcus, "Johnstown Steel Strike," 1, 63.

64. Gelotte's notice to strikers quoted in Mountjoy, "Organize the Unorganized."

65. Marcus, "Johnstown Steel Strike," 63.

66. Ibid.; Smith, "Steel Strike."

67. Foster, *Great Steel Strike*, 48.

68. Marcus, "Johnstown Steel Strike," 1.

69. Ibid.

70. Ibid.; The *Cambria Tribune* was founded by James Swank in 1853 and became the *Johnstown Tribune* in 1854. The *Johnstown Democrat* was founded in 1863. The Tribune Publishing Company bought the *Democrat* in 1934 and continued to publish both papers until 1952, when it merged the publications together.

71. Marcus, "Johnstown Steel Strike," 1.

72. Charles Weidner interview by Augusta Betzwieser, "In the Age of Steel"; Wallace, "Character."

73. Whittle, *Johnstown, Part 1*, 58; "Bethlehem Takes Midvale," *Iron Age* 111, no. 14 (April 5, 1923): 981.

74. Hessen, *Steel Titan*, 249.

75. Ibid.

76. Ibid.

77. Gertrude Fox interview by Mindy Small, "In the Age of Steel."

78. Assad, Frassinelli, Venditta and Whelan, *Forging America*, 35.

79. Ibid.

80. Warren, *Bethlehem Steel*, 52.

81. Strohmeyer, *Crisis*, 83.

82. Ibid.

83. Hessen, *Steel Titan*, 250.

84. "Production of Steel by Countries," AISI Annual Statistical Reports printed in *Iron Age* 121, no. 1 (January 5, 1933): 145; Reutter, *Sparrows Point*, 24.

85. Ibid.; Schwab speech printed in "Steel Institute to Broaden Scope as Leaders Assert Faith in Future," *Iron Age* 120, no. 21 (May 26, 1932): 173–76.

86. Assad, Frassinelli, Venditta and Whelan, *Forging America*, 86; Hessen, *Steel Titan*, 250.

87. Ibid.

Chapter 3

88. Glosser Brothers Department Store advertisement, *Tribune*, March 16, 1936; "Chaplin's Latest Picture Coming," *Tribune*, March 16, 1936; "France Dims Outlook by Planned Appeal to the Hague Court," *Tribune*, March 16, 1936.

89. "Heavy Rain and Melting Snow Turns City into Vast Lagoon," *Tribune*, March 14, 1907; Toner, "Floods," 27.

90. Toner, "Floods," 27.

91. Ibid.

92. Ibid.

93. "Streams on Rampage Throughout District as Rain Pours Down," *Tribune*, March 17, 1937; "No Flood Tomorrow, Says Weatherman," *Tribune*, March 16, 1937; Custer, "Document," 347.

94. "Incline Plane was Means of Taking Many to Safety," *Tribune*, April 1, 1936; Toner, "Floods," 28.

95. "Incline Plane," *Tribune*.

96. "15 Found Dead," *Tribune*, March 20, 1936; "25 Die in Flood," *Tribune*; "Spent Night of Flood atop Ten-Acre Bridge," *Tribune*, March 25, 1936.

97. "Flood Sidelights," *Tribune*, March 30, 1936; "Earle Throws Police Forces into Districts," *Tribune*, March 20, 1936; "Thrills Feature Rescues," *Tribune*, March 20, 1936.

98. Whittle, *Johnstown, Part 1*, 238.

99. "Earle Throws," *Tribune*; Toner, "Floods," 27–28.

100. Whittle, *Johnstown, Part 1*, 238–39.

101. "Interesting Sidelights of the Flood," *Tribune*, March 25, 1936.

102. Mike Wolfe, interview by Conrad Suppes, "1977 Oral History Project."

103. "Dams are Safe, States Shields," *Tribune*, March 20, 1936.

104. "Escaping Flood Harder at 90 Than When Aged 43," *Tribune*, March 31, 1936.

105. "25 Die in Flood," *Tribune*; "City's Second Major Deluge," *Tribune*; "Loss to Business is Heavy," *Tribune*, March 20, 1936; Wallace, "Character."

106. "Earle Throws," *Tribune*; "Johnstown Merchant Sets Loss from Deluge at $50,000," *Tribune*, March 20, 1936.

107. Violations of Free Speech and Rights of Labor, U.S. Senate Committee on Education and Labor, 14,869.

108. "25 Die in Flood," *Tribune*; Whittle, *Johnstown, Part 1*, 241.

109. "Return to Normal Shows Gain Daily in Stricken City," *Tribune*, March 24, 1936; "City in Arms at Present," *Tribune*, March 20, 1936.

110. "City in Arms," *Tribune*; "Johnstown Fights for New Foothold," *Tribune*, March 23, 1936; "Johnstown Survives Disaster," *Tribune*, March 23, 1936.

111. "Liquor Dealers to Meet Mayor, County Official," *Tribune*, March 30, 1936; "City in Arms," *Tribune.*

112. "Nine Perish When Flood Hits Districts," *Tribune*, March 20, 1936.

113. "Nine Perish," *Tribune*; "No Lack of Food in City," *Tribune*, March 20, 1936; "Fire Companies of Neighboring Towns Pump Local Cellars," *Tribune*, March 20, 1936.

114. "Memorial Hospital Pays Debt," *Tribune*, March 20, 1936; "Lenten and Friday Fasts of Catholic Church Suspended," *Tribune*, March 23, 1936.

115. "Refugee Camps at Daisytown," *Tribune*, March 23, 1936; "Locations Listed of Fuel Centers," *Tribune*, March 24, 1936; "22 American Red Cross National Staff Members Take Situation in Hand," *Tribune*, March 23, 1936.

116. "Johnstown has Escaped One of Worst Enemies," *Tribune*, March 23, 1936.

117. "A Blow Below the Belt," *Tribune*, March 31, 1936.

118. "Bethlehem Will Not Leave," *Tribune*, April 2, 1936; "Johnstown Fights," *Tribune*; "Again, Johnstown Marches On!" *Tribune*, April 4, 1936.

119. "Johnstown Wants Security," *Tribune*, March 24, 1936.

120. Willard Edwards, "Dams to Curb Flood Pledged by Roosevelt," *Chicago Tribune*, August 13, 1936.

121. House Committee on Flood Control, Hearings on Levee and Flood Walls, Ohio River Basin; "Army Engineers Study," *Tribune*; Dave Sutor, "70 Years Later, Flood Walls Still Protecting City," *Tribune-Democrat*, November 26, 2013.

122. Kozlovac, "Adventures," 4–5; Roosevelt letter to Krebs.

123. Eleanor Roosevelt Papers Digital Edition, "My Day, November 19, 1940," www2.gwu.edu.

124. Toner, "Floods," 27.

125. "Flood Sidelights," *Tribune.*

Chapter 4

126. Whittle, *Johnstown, Part 1*, 37–38, 238.

127. Ibid.

128. "LaGuardia Says Johnstown Would Be Air Raid Target," *Tribune*, December 11, 1941.

129. Carol Loomis, "The Sinking of Bethlehem Steel," *Fortune*, April 5, 2004; Whittle, *Johnstown, Part 2*, 43; "District Industries Set Production Records,"

NOTES TO PAGES 91–102

Tribune, December 30, 1944; "Local Plant Cited as Big Supplier," *Tribune*, July 19, 1945; "To Build Munitions Plant Here," *Tribune*, July 7, 1944; "Pick Coopersdale Site for Munitions Plant," *Tribune*, July 8, 1944; "Shell-Forging Plant Will Enter Production Monday," *Tribune*, February 10, 1945.
130. Excerpt of Grace's broadcast was included in the PBS documentary *Bethlehem Steel: The People Who Built America* (2004).
131. Eugene Simmers interview by Kevin Gardner, "In the Age of Steel."
132. Whittle, *Johnstown, Part 2*, 48.
133. Ibid.; "Johnstown Works, U.S. Steel Enjoys Big Year," *Tribune*, January 9, 1954; Strohmeyer, *Crisis*, 196; Loomis, "Sinking."
134. Metzgar, "Lackawanna and Johnstown," 69.
135. Frank Guidon interview by Kathy Schlegel, "In the Age of Steel."
136. Herbert Sechler interview by Kathleen Munley, "In the Age of Steel."
137. John Whitney interview by Kathy Schlegel, "In the Age of Steel."
138. Annual Reports for 1950 and 1960, Pennsylvania Department of Mines, Bituminous Division; Whittle, *Johnstown, Part 2*, 75.
139. Ibid.; Ciervo, *Always in a Hole*, 44.
140. Whittle, *Johnstown, Part 2*, 172.
141. Ibid.; Reutter, *Sparrows Point*, 420.
142. George Dancho interview by Mindy Small, "In the Age of Steel."
143. "R. Conrad Cooper Dies at 79," *New York Times*, October 2, 1982; Strohmeyer, *Crisis*, 66.
144. Rose, "Struggle over Management Rights," 461.
145. Strohmeyer, *Crisis*, 69.
146. Shils, "Goldberg," 62.
147. Rose, "Struggle over Management Rights," 461.
148. Metzgar, *Striking Steel*, 55.
149. Strohmeyer, *Crisis*, 85–86.
150. Assad, Frassinelli, Venditta and Whelan, *Forging America*, 128.
151. Ibid; Whittle, *Johnstown, Part 2*, 173–74.
152. Ibid.
153. Strohmeyer, *Crisis*, 197; Assad, Frassinelli, Venditta and Whelan, *Forging America*.
154. "Bethlehem to Cut Operations in Johnstown in Half By '77," *Tribune-Democrat*, June 13, 1973; Whittle, *Johnstown, Part 2*, 175.
155. Metzgar, "Lackawanna and Johnstown," 67.

Chapter 5

156. Larry Hudson, "Horror of Johnstown's History Becomes Reality Again," *Tribune-Democrat*, July 22, 1977.

157. Bosart and Sanders, "Flood of July 1977," 1,616–17; "Flood Controls No Match for Sudden Deluge," *Altoona Mirror* (hereafter *Mirror*), July 21, 1977.

158. Nina Kalinyak, "A Gift Twice Given," *Tribune-Democrat*, July 24, 1977.

159. Founded in 1950, the Jets captured back-to-back Eastern Hockey League titles in 1952 and 1953. The team competed in three different leagues—the International Hockey League, the Eastern Hockey League and the North American Hockey League—before folding following the 1976–77 season (Jackson, *Making of* Slap Shot, 36–40).

160. Bill Hindman interview by Conrad Suppes, "1977 Oral History Project."

161. The other six counties that were issued flood warnings were Armstrong, Butler, Clarion, Clearfield, Indiana and Jefferson (Ted Chiappelli, "Flooding Unavoidable," *Tribune-Democrat*, July 24, 1977); Janet Smetto-Mical interview by Conrad Suppes, "1977 Oral History Project."

162. Rich Horner interview by Conrad Suppes, "1977 Oral History Project"; "Former Mayor Pfuhl Dies," *Tribune-Democrat*, August 20, 2011.

163. Whittle, *Johnstown, Part 2*, 182–83; "Lee Transferring All Patients," *Tribune-Democrat*, July 23, 1977.

164. Larry Hudson, "The Flood of 1977," *Tribune-Democrat*, July 24, 1977.

165. Tom Gibb, "Over 100 Still Missing in Johnstown," *Mirror*, July 23, 1977; John McHugh, "City Flood Loss Expected to Top $100 Million," *Tribune-Democrat*, July 20, 1977; Sandra Reabuck and Cindy Burkett, "City Waters Began to Build," *Tribune-Democrat*, July 20, 1977.

166. Hudson, "Horror."

167. George Fattman interview by Conrad Suppes, "1977 Oral History Project."

168. Chiappelli, "Flooding Unavoidable."

169. "Clean-up Work Begins in Wake of Downpour," *Mirror*, July 21, 1977; Tom Gibb, "In the Aftermath," *Mirror*, July 21, 1977; Lee Hospital reopened on August 1. The only part of Conemaugh Hospital that flooded was a tunnel under Franklin Street that connected the hospital to a nursing school affiliated with the facility. (Whittle, *Johnstown, Part 2*, 188.)

170. McHugh, "City Flood Loss."

171. Larry Hudson, "Home Scarred but Victorious," *Tribune-Democrat*, August 1, 1977.

172. Federal Guidelines for Dam Safety, Federal Emergency Management Agency, Association of State Dam Safety.

173. Ed Cernic interview by Conrad Suppes, "1977 Oral History Project"; David Hurst, "1977 Flood," *Tribune-Democrat*, July 16, 2017.

174. Carol Burns interview by Conrad Suppes, "1977 Oral History Project."

175. Ibid; Hudson, "Flood of 1977"; John McHugh and William Black, "Death Toll Stands at 55," *Tribune-Democrat*, July 22, 1977.

176. "The Men Behind the Relief Effort," WTAJ, July 18, 2017; "River Backed Up, But How Much?" *Tribune-Democrat*, April 1, 1979.

177. Ibid; Reclamation Consequence Estimation Methodology, U.S. Department of the Interior Bureau of Reclamation.

178. "River Backed Up," *Tribune-Democrat*.

179. Whittle, *Johnstown, Part 2*, 186; Kris Jenkins, "Blair Citizens Rally to Help Neighbors," *Mirror*, July 22, 1977; Cernic interview, "1977 Oral History Project."

180. Federal Guidelines for Dam Safety, Federal Emergency Management Agency, Association of State Dam Safety.

181. "City Faces Recovery Task for Third Time," *Tribune-Democrat*, July 20, 1977; Majumdar, *Natural and Technological Disasters*, 44.

182. Ted Potts, "7,000 People Displaced by Flood," *Tribune-Democrat*, July 26, 1977; Harold Martin, "Conemaugh River Valley Warned of New Flood Threats," *Mirror*, July 25, 1977.

183. "Flood Controls," *Mirror*, 10; "I Couldn't Hold onto My Mother," *Mirror*, July 21, 1977.

184. "Flood Controls," *Mirror*, 10; Chiappelli, "Flooding Unavoidable"; Hudson, "Horror"; "Couldn't Hold," *Mirror*; Scott MacLeod, "Devastation Widespread," *Mirror*, July 21, 1977.

185. Whittle, *Johnstown, Part 2*, 190.

186. "City Opens Lifeline," *Mirror*, July 21, 1977; "Rail Freight Traffic on Move Again," *Mirror*, July 25, 1977.

187. "Resumes Service," *Mirror*, July 25, 1977.

188. Whittle, *Johnstown, Part 2*, 188; John McHugh, "Flood Relief Effort Grows," *Tribune-Democrat*, July 23, 1977; "Plane Took 1,500 to Hilltop," *Tribune-Democrat*, July 24, 1977.

189. Joyce Murtha interview by Conrad Suppes, "1977 Oral History Project."

190. Gibb, "Over 100 Still Missing"; James Siehl, "Richland School Overflowing as Homeless Pour In," *Tribune-Democrat*, July 24, 1977; Potts, "7,000 People Displaced"; Author interview with Christian Oravec, April 1, 2013.

191. "Survivor is Now Afraid of Water," *Mirror*, July 21, 1977, 10.
192. Martin, "Conemaugh River Valley Warned"; "Grim Vigil Continues," *Mirror*, July 23, 1977.
193. Hudson, "Flood of 1977"; Scott MacLeod, "Death Toll, Damage Estimates Escalate," *Mirror*, July 23, 1977; Kris Jenkins, "Senior Citizens Set Up Lifeline," *Mirror*, July 25, 1977.
194. "Meeting at White House," *Mirror*, July 21, 1977.
195. McHugh, "Flood Relief Effort"; Hudson, "Flood of 1977"; MacLeod, "Death Toll"; "Contractors to Hear Repair Work Details," *Mirror*, July 28, 1977; Cernic interview, "1977 Oral History Project."
196. Brad Clemenson, "Rescue Squads at Work," *Tribune-Democrat*, July 24, 1977.
197. Hurst, "1977 Flood."
198. Robert Sefick, "End of the World in Windber," *Tribune-Democrat*, July 24, 1977.
199. Kay Stephens, "Huge Clean-up Effort Begins," *Mirror*, July 23, 1977; Others also noted that the sky appeared to be green before sunset on the night of the flood. ("Flooding is Over," *Mirror*, July 23, 1977.)
200. Clifton Crosbie, "Windber Loss at $30 Million," *Tribune-Democrat*, July 25, 1977; "Roundup of Flood Damage," *Tribune-Democrat*, July 25, 1977; Sefick, "End of the World"; Hurst, "1977 Flood."
201. "Roundup," *Tribune-Democrat*; Hudson, "Horror"; Hurst, "1977 Flood."
202. Gibb, "In the Aftermath," 28; "Weeks of Clean-Up," *Mirror*, July 21, 1977.
203. Joseph Kasprzyk, "Mainline Area Losses Estimated at $4.5 Million," *Tribune-Democrat*, July 23, 1977; "Damage Extensive in Sections of Portage," *Mirror*, July 21, 1977.
204. Martin, "Conemaugh River Valley Warned."
205. Sidney Goldblatt interview by Conrad Suppes, "1977 Oral History Project."
206. Hindman interview, "1977 Oral History Project."
207. Kalinyak, "Gift Twice Given."
208. Norm Verhovsek interview by Conrad Suppes, "1977 Oral History Project."
209. Hurst, "1977 Flood."
210. Ibid.
211. "Couldn't Hold," *Mirror*.
212. "Survivor 'Still Loves Johnstown,'" *Mirror*, July 23, 1977.
213. Ibid.
214. McHugh, "Flood Relief Effort."
215. Ibid.

216. Hindman interview, "1977 Oral History Project."

217. Gary Henderson interview by Conrad Suppes, "1977 Oral History Project."

218. "Remembers Other Floods," *Mirror*, July 21, 1977.

219. Scott MacLeod, "Clouds of Uncertainty Hang Over City," *Mirror*, July 22, 1977; "Centers Opening," *Mirror*, July 25, 1977.

220. MacLeod, "Devastation Widespread"; Francis Ozog interview by Conrad Suppes, "1977 Oral History Project."

221. Hudson, "Flood of 1977"; "Flood Controls," *Mirror*; "Pilot Stunned by Extent of Damage," *Mirror*, July 21, 1977; MacLeod, "Devastation Widespread."

222. MacLeod, "Devastation Widespread"; Hudson, "Flood of 1977."

223. Horner interview, "1977 Oral History Project."

224. "Cambria County Getting Too Much Relief Clothing," *Mirror*, July 25, 1977; Crosbie, "Windber Loss."

225. "'Til We're Normal Again, Here's How News Gets Out," *Tribune-Democrat*, July 22, 1977; "Restoring Service," *Mirror*, July 25, 1977; The *Tribune-Democrat* actually printed and distributed two editions of the paper on July 20, 1977. When the flood hit, that day's edition of the paper was coming off the press. Those papers were sent out, and then an updated edition with information on the flood was printed at the *Tribune-Review* offices in Greensburg. (Norm Verhovsek and George Fattman interview by Conrad Suppes, "1977 Oral History Project.")

226. Dick Mayer interview by Conrad Suppes, "1977 Oral History Project."

227. Ibid.

228. Fattman interview, "1977 Oral History Project."

229. Susan Brett interview by Conrad Suppes, "1977 Oral History Project."

230. Mayer interview, "1977 Oral History Project"; Gerald Swatsworth and George Fattman interview by Conrad Suppes, "1977 Oral History Project."

231. Caram Abood interview by Conrad Suppes, "1977 Oral History Project."

232. Mike Faher, "Late Priest Remembered," *Tribune-Democrat*, July 20, 2007; Brett interview, "1977 Oral History Project"; Slavik was killed in a car accident in February 1982.

233. "Spirit of Recovery," *Tribune-Democrat*, August 1, 1977.

234. Toby Sweeney, "Looters Hit Business," *Tribune-Democrat*, July 20, 1977; Gibb, "In the Aftermath"; "Clean-up Work Begins," *Mirror*.

235. "Couldn't Hold," *Mirror*; "Clean-up Work Begins," *Mirror*; Cernic interview, "1977 Oral History Project."

236. McHugh, "Flood Relief Effort"; McHugh, "City Flood Loss"; Gibb, "Over 100 Still Missing."

237. "FBI Director Kelly Backs Pfuhl Order to Shoot Looters," *Somerset Daily American*, July 27, 1977.

238. "Two-Gun Herb," *Philadelphia Daily News*, July 29, 1977.

239. Nancy Maneely, "Residents Say Police Knew," *Tribune-Democrat*, July 24, 1977.

240. Ibid.

241. Ibid.

242. "Contractors," *Mirror*; Kris Jenkins, "Blair Citizens Rally"; "Carrying Water to Tanneryville," *Tribune-Democrat*, August 1, 1977.

243. "Contractors," *Mirror*.

244. Rothstein, "Four Years."

245. Pohla Smith, "Forecaster Urges Flood Alert Plan," *Mirror*, July 23, 1977.

246. Ibid.

247. Hindman interview, "1977 Oral History Project"; Roosevelt letter to Krebs; Federal Guidelines for Dam Safety, Federal Emergency Management Agency, Association of State Dam Safety.

248. "Flood Controls," *Mirror*.

249. "Spirit of Recovery," *Tribune-Democrat*.

Chapter 6

250. "Bethlehem Plant Damaged Extensively," *Tribune-Democrat*, July 20, 1977.

251. "Officials to Meet on Steel Plant's Future," *Tribune-Democrat*, August 1, 1977.

252. Strohmeyer, *Crisis*, 198.

253. "Bethlehem to Close Coke Battery," *Tribune-Democrat*, January 25, 1982; Serrin, "Steel Industry Woes."

254. Whittle, *Johnstown, Part 2*, 172.

255. Assad, Frassinelli, Venditta and Whelan, *Forging America*, 125.

256. Strohmeyer, *Crisis*, 198; Bill Toland, "In Desperate 1983, There Was Nowhere for Pittsburgh's Economy to Go But Up," *Pittsburgh Post-Gazette*, December 23, 2012.

257. Scott Kraus, "Former Bethlehem Steel CEO Donald Trautlein Dies," *Allentown Morning Call*, July 8, 2010.

258. Strohmeyer, *Crisis*, 198, 199–200, 203.

259. Ibid.

260. Kraus, "Bethlehem CEO Dies."

261. Strohmeyer, *Crisis*, 201–4.

262. Ibid.

263. Ken Kunsman, "Steel Finalizes Sale of Division in Johnstown," *Allentown Morning Call*, July 7, 1993; The wire mill continues to operate in the city today. The Franklin plant operated under two different owners until its closure in 1997. ("Indian Firm Shows Interest in Bethlehem Division," *Baltimore Sun*, November 12, 1992.)

264. Peter Behr, "Bethlehem Steel Files for Bankruptcy," *Washington Post*, October 16, 2001.

265. *Bethlehem Steel: The People Who Built America*, aired on PBS in 2004.

266. Loomis, "Sinking."

Epilogue

267. Donato Zucco interview by Conrad Suppes, "1977 Oral History Project."

268. Crandall, *Continuing Decline*, 122–23.

269. "Best Small Places for Businesses and Careers," *Forbes Magazine*, 2019.

270. Ibid.

271. Author interview with Randy Frye, August 3, 2020.

272. Murtha helped to bring $90 million in low-interest loans to small businesses in the community.

273. Jason Zengerle, "Will Murtha's Town Die with Him," *New Republic*, February 9, 2010.

274. Leonardo DRS is a defense contractor with plants across the United States. Its division in Johnstown is called Laurel Technologies, and it produces naval electronics equipment. Concurrent Technologies is an engineering and manufacturing research and development company. Lockheed Martin AeroParts is part of Lockheed Martin's global holdings. Its Johnstown division produces aeronautical equipment for the U.S. military. (Interview with Frye.)

275. Dave Sutor, "After the Flood," *Tribune-Democrat*, July 18, 2020.

276. Ibid.

277. Estimates of population.

278. Author interview with Dale Falcinelli, August 3, 2020.

279. "How Wind Creek Bethlehem Helped Pa. Hit a New Record," *Associated Press*, January 17, 2020; Anthony Salamone, Jon Harris, and Tom Shortell, "Will the Lehigh Valley Became a Hip Hub in the Next Decade?" *Allentown Morning Call*, January 2, 2020.

280. Interview with Falcinelli.

281. Interview with Frye; Some steel companies had successfully transitioned into the "mini-mill" business, producing steel rolled from electric steel furnaces at much lower production costs. (Assad, Frassinelli, Venditta and Whelan, *Forging America*, 125.)

282. A transcript of Elder's speech was reproduced in a publication documenting the city's 1900 centennial. It is archived at the Johnstown Area Heritage Association.

283. "The Private Strategy of Bethlehem Steel," *Fortune*, April 1962; Reutter, *Sparrows Point*, 420–21.

284. Ibid.

285. *Bethlehem Steel*, PBS.

286. Ibid.

287. Shugerman, "Floodgates," 333.

BIBLIOGRAPHY

Oral History Collections

"In the Age of Steel: Oral Histories from Bethlehem, Pennsylvania." Collection housed at Lehigh University, Bethlehem, PA.

"1977 Oral History Project: Documenting the Stories of the July 20, 1977 Flood." Collection housed at Johnstown Area Heritage Association, Johnstown, PA.

Published Government Documents

Alien Registration Act of 1940 (i.e. Smith Act). Seventy-Sixth United States Congress, 3rd session, ch. 439, 54 stat. 670, 18 U.S.C. § 2385.

Annual Reports for 1950 and 1960. Pennsylvania Department of Mines, Bituminous Division.

County Health Rankings Report—Pennsylvania. University of Wisconsin Population Health Institute, 2020.

Federal Guidelines for Dam Safety. Federal Emergency Management Agency, Association of State Dam Safety, 2004.

Hearings on Levee and Flood Walls, Ohio River Basin. House Committee on Flood Control, Seventy-Fifth United States Congress, 1st session, June 7–11 and June 15–18, 1937.

A History of Appalachian Coal Mines in Legal Problems of Coal Mine Reclamation. Washington, D.C.: U.S. Government Printing Office, 1972.

Immigrants in Industries, Part 2. Immigration Commission, Sixty-First United States Congress, 2nd sess., 1911, S. doc 633.

Pennsylvania Bureau of Mines Annual Report, 1900.

Reclamation Consequence Estimation Methodology. U.S. Department of the Interior Bureau of Reclamation, 2014.

Toner, Peter. "Floods of Johnstown." Federal Writers Project, Works Progress Administration of the Commonwealth of Pennsylvania, 1939.

Violations of Free Speech and Rights of Labor. U.S. Senate Committee on Education and Labor. Mishawaka, IN: Palala Press, 2015.

Wallace, Kim. "The Character of a Steel Mill City." Historic American Buildings Survey, National Park Service, Washington, D.C., 1989.

Newspapers

Allentown Morning Call (Allentown, PA)
Altoona Mirror (Altoona, PA)
Baltimore Sun (Baltimore, MD)
Chicago Tribune (Chicago, IL)
Johnstown Tribune (Johnstown, PA)
Johnstown Tribune-Democrat (Johnstown, PA)
Johnstown Weekly Democrat (Johnstown, PA)
New York Times (New York, NY)
Philadelphia Daily News (Philadelphia, PA)
Pittsburgh Post-Gazette (Pittsburgh, PA)
Somerset Daily American (Somerset, PA)
Three Rivers Tribune (Pittsburgh, PA)
Washington Post (Washington, D.C.)

Magazines

Air Force Magazine (Arlington, VA)
Forbes Magazine (Jersey City, NJ)
Harper's Weekly (New York, NY)
Huffington Post (New York, NY)
New Republic (New York, NY)
TIME (New York, NY)

Secondary Sources

Assad, Matt, Mike Frassinelli, David Venditta and Frank Whelan. *Forging America*. Allentown, PA.: Morning Call, 2010.

Berger, Karl, and William Glosser. *Johnstown: The Story of a Unique Valley*. Johnstown, PA: Johnstown Flood Museum, 1985.

Bosart, Lance, and Frederick Sanders. "The Johnstown Flood of July 1977." *Journal of the Atmospheric Sciences* 38, no. 8 (1981): 1,616–642.

Ciervo, Arthur. *Always in a Hole*. Camp Hill, PA: Suburban Press, 1996.

Coleman, Neil. *Johnstown's Flood of 1889*. New York: Springer, 2019.

Crandall, Robert. *The Continuing Decline of Manufacturing in the Rust Belt*. Washington, D.C.: Brookings Institute, 1993.

Custer, Dale. "A Document on the Second Johnstown Flood." *Pennsylvania History* 30, no. 3 (1963): 347–54.

Dickens, Charles. *American Notes for General Circulation*. London: Chapman & Hall, 1842.

Donnelly, Lu. *Buildings of Pennsylvania*. Charlottesville: University of Virginia Press, 2010.

Foster, William. *The Great Steel Strike and Its Lessons*. New York: B.W. Huebsch, 1920.

Frank, Walter. "The Cause of the Johnstown Flood." *Civil Engineering* 58, no. 5 (May 1988): 63–66.

Harper, Eugene. "Town Development in Early Western Pennsylvania." *Western Pennsylvania* 71, no. 1 (January 1988): 3–26.

Hessen, Robert. *Steel Titan*. New York: Oxford University Press, 1975.

Hutcheson, Edwin. *Floods of Johnstown: 1889–1936–1977*. Johnstown, PA: Cambria County Tourist Council, 1989.

Jackson, Jonathon. *The Making of* Slap Shot. Mississauga, ON: Wiley, 2010.

Jenkins, Phillip. *Hoods and Shirts*. Chapel Hill: University of North Carolina Press, 1997.

Kozlovac, Joseph. "Adventures in Flood Control: The Johnstown, Pennsylvania Story." *Urban Areas as Environments* 14, no. 4 (April 19, 1995): 1–12.

Majumdar, Shyamal. *Natural and Technological Disasters*. University Park: Pennsylvania Academy of Science, 1992.

Marcus, Irwin. "The Johnstown Steel Strike of 1919: The Struggle for Unionism and Civil Liberties." *Pennsylvania History* 63, no. 1 (January 1996): 96–118.

McCullough, David. *The Johnstown Flood*. New York: Simon & Schuster, 1968.

McDevitt, Cody. *Banished from Johnstown*. Charleston, SC: The History Press, 2020.

Metzgar, Jack. "Lackawanna and Johnstown." *Labor Research Review* 1, no. 2 (1983): 66–72.

———. *Striking Steel*. Philadelphia, PA: Temple University Press, 2000.

Mountjoy, Eileen. "To Organize the Unorganized." Indiana University of Pennsylvania Special Collections and Archives, IUP Libraries.

Nasaw, David. *Andrew Carnegie*. New York: Penguin Press, 2006.

Reutter, Mark. *Sparrows Point: Making Steel*. New York: Summit Books, 1988.

Roker, Al. *Ruthless Tide*. New York: Harper Collins, 2018.

Rose, J.D. "The Struggle Over Management Rights at U.S. Steel, 1946–1960." *Business History Review* 72 (Fall 1998): 446–77.

Shappee, Nathan. "A History of Johnstown and the Great Flood of 1889." PhD dissertation, University of Pittsburgh, 1940.

Sherman, Richard. "Johnstown v. the Negro." *Pennsylvania History* 30, no. 4 (October 1963): 454–64.

Shils, Edward. "Arthur Goldberg." *Monthly Labor Review* 120, no. 1 (January 1997): 56–71.

Shugerman, Jed. "The Floodgates of Strict Liability: Bursting Reservoirs and the Adoption of Fletcher v. Rylands in the Gilded Age." *Yale Law Journal* 110, no. 2 (November 2000): 333–77.

Smith, Robert. "The Steel Strike in Johnstown, Pennsylvania, 1919." Master's thesis, Indiana University of Pennsylvania, 1982.

Strohmeyer, John. *Crisis in Bethlehem*. Bethesda, MD: Adler & Adler, 1986.

Tarr, David. *Issues in US-EC Trade Relations*. Chicago, IL: University of Chicago Press, 1988.

Warren, Kenneth. *Bethlehem Steel*. Pittsburgh, PA: University of Pittsburgh Press, 2009.

Warrian, Peter. *A Profile of the Steel Industry*. New York: Business Expert Press, 2016.

Whittle, Randy. *Johnstown, Pennsylvania: A History, Part 1, 1895–1936*. Charleston, SC: The History Press, 2005.

———. *Johnstown, Pennsylvania: A History, Part 2, 1937–1980*. Charleston, SC: The History Press, 2007.

INDEX

F

Falcinelli, Dale 148, 149, 153
Family Store 129
Fattman, George 107, 127
Federal Aviation Agency 132
Federal Bureau of Investigation 130
Federal Emergency Management Agency 110
Federal Reserve 66
Fenn, Anna 39, 40
Fenn, John 39
Ferndale 75, 107
Ferndale-Dale School District 105
First Baptist Church of Johnstown 58
First United Brethren Church of Johnstown 69
Flood-Free Johnstown Committee 16, 86
Foster, William 55, 57
Fox, Gertrude 61
Foy, Lewis 101, 102, 135, 136, 137
Franklin 49, 132, 135, 139, 140
Franklin Street Bridge 73
Frazier Refrigeration 126
Freville, Harvey 126, 129
Frick, Henry Clay 28
Fritz, Herbert 80
Frye, Randy 145, 146, 150
Fulton, John 32, 33, 34

G

Gallagher, Dan 70
Galliker, Louis 147
Galliker's Dairy 108, 124
Garvey, Eugene 58
Gatehouse, Ron 107

Gautier Company 32, 140
Gautier, Josiah 32
Geistown 124
Gelotte, Dominick 56
General Telephone Company 126
Genovese, Dominick 105
Gleason, Robert 147
global pandemic 144, 147, 148
Glosser Brothers Department Store 67, 73, 129
Glosser, Daniel 147
Goldberg, Arthur 97
Goldblatt, Sidney 122
Gompers, Samuel 55
Gooseneck Island 122
Govekar, Joseph 124
Grace, Eugene 60, 66, 91, 92, 95, 99, 100
Grand Army Memorial Hall 80
Grandview Cemetery 42
Great Depression 52, 66
Greensburg 127
Grub, Billy 70
Grubbtown 49
Guffey, Joseph 85
Guidon, Frank 93
Gulf of Mexico 22

H

Harrisburg 116, 118, 119, 149
Harrison, James 21
Harrison, William Henry 23
Hastings, Daniel 42
Haynes Street Suspension Bridge 68
Heinz, John 100, 118, 135
Hellman, Connie 125
Hellman, Maria 125
Hellman, Ron 125

ABOUT THE AUTHOR

 Pat Farabaugh is a professor of communications at Saint Francis University in Loretto, Pennsylvania. He has also taught at Indiana University of Pennsylvania and Penn State University. He earned his PhD in political and cultural communications from Penn State. His previous books include *Carl McIntire's Crusade Against the Fairness Doctrine* and *An Unbreakable Bond: The Brotherhood of Maurice Stokes and Jack Twyman*. He is also a contributing author to *American Sports: A History of Icons, Idols, and Ideas*. He and his wife, Jenna, live in Indiana, Pennsylvania.

Visit us at
www.historypress.com
....